$95

Polymer Modification

Polymer Modification

Edited by

Graham Swift

Rhome and Haas Research
Spring House, Pennsylvania

Charles E. Carraher, Jr.

Florida Atlantic University
Boca Raton, Florida, and
Florida Center for Environmental Studies
Palm Beach Gardens, Florida

and

Christopher N. Bowman

University of Colorado at Boulder
Boulder, Colorado

PLENUM PRESS • NEW YORK AND LONDON

Library of Congress Cataloging-in-Publication Data

Polymer modification / edited by Graham Swift, Charles E. Carraher, Jr., and Christopher N. Bowman.
p. cm.
Papers presented at a symposium held at the American Chemical Society Meeting in Orlando, Fla., Aug. 1996.
Includes bibliographical references and index.
ISBN 0-306-45714-8
1. Polymers--Congresses. I. Swift, Graham, 1939- . II. Carraher, Charles E. III. Bowman, Christopher N., 1967- . IV. American Chemical Society. Meeting (212th : 1996 : Orlando, Fla.)
TP1081.P64 1998
668.9--dc21 97-35159
CIP

ISBN 0-306-45714-8

A Division of Plenum Publishing Corporation
233 Spring Street, New York, N.Y. 10013

http://www.plenum.com

10 9 8 7 6 5 4 3 2 1

Printed in the United States of America

PREFACE

Polymer modifications represent a valuable synthetic approach to unique polymer compositions, structure, and properties not readily available by the direct polymerization of monomers. Modified polymeric products already exist in the commercial world (modified celluloses, for example) so the approach is not new. However, it is an interesting and challenging opportunity to develop new materials for a variety of specialty applications using the "chemistry on polymers" approach.

This book contains papers presented at the symposium on Polymer Modification held at the National American Chemical Society Meeting in Orlando, Florida, August, 1996. The chemistry presented is broad ranging, and includes grafting and chemical oxidation reactions, and many other chemical modifications.

Hopefully, the book will be both a resource and an inspiration for the reader to develop new opportunities for his or her particular applications.

CONTENTS

SURFACE MODIFICATIONS

REACTIONS WITH VINYL POLYMERS

INORGANIC-CONTAINING AND SHAPED POLYMERS

Polymer Modification

SURFACE MODIFICATIONS

THE PREPARATION OF METHYL METHACRYLATE/ METHACRYLIC ANHYDRIDE COPOLYMERS FROM PMMA AND DIALKYL AMINES VIA REACTIVE EXTRUSION

Michael P. Hallden-Abberton

Rohm and Haas Co.,
Corporate Exploratory Research,
P.O. Box 219,
Bristol, Pennsylvania 19007-0219

INTRODUCTION

Although the imidization reaction of PMMA has been studied for many years[1-4] recently, the study of this reaction by reactive extrusion processes has led to some interesting and unexpected results.[5-7,11] Specifically, the first step of this reaction is now believed to be due to an unusual alkyl oxygen cleavage reaction resulting in the formation of methacrylic acid and methacrylic anhydrides.[7] In an attempt to corroborate this mechanism, we have conducted some model reactions using dimethyl amine and PMMA in a twin screw, counter-rotating, non-intermeshing extruder. According to the proposed mechanism,[7] the reaction products should be derived from these alkyl-oxygen cleavage products. However, by use of a dialkyl amine, the intermediate products should be prevented from cyclization to imide products, and therefore should be trapped at an intermediate stage. This should allow some evidence of the reaction pathway to be observed. Also, analysis of the gaseous products should be indicative of the reaction pathway.

REACTIVE EXTRUSION PROCEDURE WITH DIMETHYLAMINE

The test reaction was conducted according to the following procedure.[10] A granular PMMA sample (Mw = 200,000) was added at a rate of 30 g/min. to the feed zone of a 20 mm, counter-rotating, tangential, twin-screw extruder operating at 250 rpm. The polymer was conveyed through a feed zone 10.2 diameters long, operated at a set temperature of 200°C, and over a melt blister 3 cm long, into a pressurized reaction zone 36.9 diameters long, operated at a set temperature of 302°C, into which dimethyl amine (Matheson Chem. Co.) was introduced by pumping at a rate of 5.2 cm^3/min (3.3 g/min) at 2515 kPa at a point 5.7 diameters downstream from the end of the melt blister. The excess amine and gaseous

Polymer Modification
Edited by Swift *et al.*, Plenum Press, New York, 1997

by-products were removed through a pressure regulated valve set to 2515 kPa at a point 27.0 diameters downstream from the end of the melt blister. The resulting polymer was then conveyed over a second melt blister 10.7 cm long and into a 21.9 diameters-long devolatilization zone at 288-302°C and 8.5 kPa pressure. The polymer at the end of the devolatilization zone was conveyed into a single screw, pumped out of a melt die, and was briefly cooled in a water bath prior to being pelletized and collected.

ANALYSIS

The product polymer (sample 1) had a Vicat softening temperature of 136.0°C (the Vicat penetration temperatures are *ca.* 6°C above the T_g of the polymer), and was analyzed for % nitrogen, and for mole % MMA, anhydride (ANH), methacrylic acid (MAA), and N,N' dimethyl methacrylamide (DMMAm) by FTIR, and carbon 13 NMR. The processing conditions and some results are shown in Table 1 and the analytical results in Table 2 (both mole % and weight %). All mole percent values are based on the functional carbonyl group analyzed divided by total carbonyl groups (to avoid confusion due to anhydride containing two carbonyl groups). In this manner, the total number of moles derived from the acrylic starting material does not change with the degree of conversion to anhydride.

In this and all other examples, the Vicat softening point is determined by ASTM DI-525-70 and reported in degrees Centigrade. The total titratable acidic functionality in the copolymethacrylanhydrides was determined by titration of a 0.04 - 0.10 g sample of polymer dissolved in 75 cc DMSO / 50 cc H_2O (sample weight adjusted to give about 0.5 mmole titer). To this stirred solution, an excess of 0.1 N NaOH is added (ca. 0.9-1.0 mmole, total) at a rate of 1.0 cc/min at room temperature. This solution is then back-titrated with 0.1 N HCl after adding phenolphthalein indicator. The total acid level is taken as the difference between the mmol of NaOH added and the mmol of HCl to neutrality per gram of sample. A second value was calculated from the mmol of HCL required to go from the first potentiometric transition ($OH^- \rightarrow H_2O$) to the second transition ($COO^- \rightarrow COOH$). The two titers thus obtained were always within 7 % of each other. Titrations were followed colorimetrically and potentiometrically, and were performed on a Metrohm Herisaw Potentiograph E536 with a Fisher Scientific pH probe (No. 1363990), or an automatic Metrohm Titroprocessor 636 instrument.

Fourier transform infrared spectra were obtained on films cast from DMSO solution (on KBR) or in solution (DMSO) on a Digilab FTS-15/90 FTIR instrument (Figure 1).

Carbon 13 spectra were run on a Varian XL-400 (400 MHz) instrument in d5 pyridine at 24°C, using a 15 second pulse delay for quantitative analysis purposes.

The PMMA/anhydride copolymer produced in sample 1 was titrated as previously described, and showed 3.57 ± 0.07 mmol/g of titratable acidic functionality. Under these titration conditions all anhydride is hydrolyzed to carboxylate and back titrated to carboxylic acid. Both NMR and IR analysis show about 27 weight % acidic functionality which is almost exclusively in the anhydride form. The NMR and IR measurements predict 3.53 and 3.48 mmol/g acidity, in good agreement with the measured titration value of 3.57 ± 0.07 mmol/g.

In a similar manner, samples 2-6 were prepared by keeping the feed polymer conditions essentially constant and varying only the amount and pressure of the dimethylamine introduced. Table 2 shows the compositional analysis of other samples which also show good agreement between the IR, % N (amide), and titration data. The IR analysis could detect about 1 % of a component or more.

RESULTS

The results of analysis of the samples prepared via reactive extrusion of PMMA with dimethylamine are shown in Tables 1 and 2, and Figure 1. It is apparent from these results that the main product of the reaction was methacrylic anhydride, confirming the hypothesis that alkyl-oxygen cleavage was involved in the reaction of PMMA with amines (at least with dimethylamine). However, additional subtleties of the reaction are apparent from a more detailed analysis of the reaction products vs. degree of conversion and reaction conditions.

Reaction Conditions vs. Product

Overall, samples 1-6 show that the reaction of dimethylamine with PMMA is facile, results in few side-products (MAA or amide) up to about 70 mole % conversion, and requires about 1:1 dimethylamine:MMA stoichiometery. As conversion approaches the Flory limit (ca. 86 mole %), more acid (MAA) is seen, and finally at high excess amine levels and pressures, a significant amount of N,N′-dimethyl methacrylamide is seen (sample 6). Other experiments at DMA/MMA ratios of 0.25-0.83 and DMA pressures of up to 4240 KPa did not produce any significant differences in anhydride or amide levels. The FTIR spectra are shown in Figure 1. Note that the anhydride has a coupled carbonyl stretch (asym=1802 cm^{-1}, sym=1759 cm^{-1}). This agrees well with previous literature (1800, 1760 cm^{-1})[8] and (1800, 1758 cm^{-1}).[9]

Table 1. Reaction Conditions vs. Vicat, % Nitrogen and Titration Results

Sample Number	Me_2NH Rate (cc/m g/m)	Me_2NH Pressure (kPa/ psig)	eq. Me_2NH/ eq. MMA (Ratio)	Vicat Temperature (°C)	% Nitrogen (%N)	mmol/g total titratable acidity
1	5.20/ 3.35	2515/ 350	0.25	136.0	0.01	3.57
2	10.5/ 6.77	2515/ 350	0.50	143.5	0.03	6.69
3	17.3/ 11.2	2515/ 350	0.83	155.3	0.10	9.92
4	28.5/ 18.4	2549/ 355	1.32	158.9	0.19	11.25
5	40.0/ 25.8	2480/ 345	1.85	158.8	0.18	10.96
6	40.0/ 25.8	3858/ 545	1.85	166.2	0.95	11.00

Table 2. Polymer Composition vs. Sample (Mole % and Weight %)

Sam-ple #	Test Type	Mole %				Weight %				Total Acid Pred/ Meas[1]
		MMA (%)	ANH (%)	MAA (%)	Amide (%)	MMA (%)	ANH (%)	MAA (%)	Amide (%)	
1	FTIR	72.8	27.2	n.d.	n.d.	67.3	32.7	n.d.	n.d.	3.53
	13C:	73.2	26.1	0.7	n.d.	67.8	31.4	0.8	n.d.	3.48/
	% N:				0.1				0.1	
	Titr:									3.57
2	FTIR	43.3	56.7	n.d.	n.d.	49.8	50.2	n.d.	n.d.	6.51/
	% N:				0.2				0.2	
	Titr:									6.69
3	FTIR	19.4	75.1	4.7.	n.d.	23.6	70.4	4.9	n.d.	9.70/
	% N:				0.6				0.8	
	Titr:									9.92
4	FTIR	6.1	87.3	5.8.	trace	7.7	84.9	6.3.	trace	11.7/
	% N:				1.1				1.5	
	Titr:									11.3
6	FTIR	1.9	77.7	12.5	7.9	2.3	73.5	13.2	11.0	11.1/
	% N:				5.5				7.7	
	Titr:									11.0

[1] Total titratable acidity (MAA+ANH in meq/g) predicted from test vs. measured value by titration.

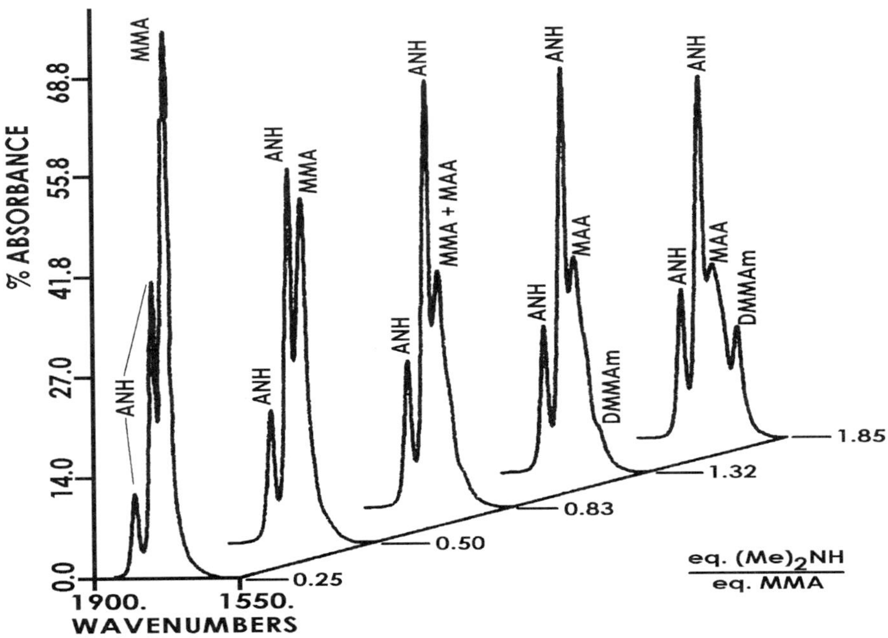

Figure 1. FTIR of carbonyl region of product polymers vs. eq. $(Me)_2NH$/eq. MMA.

DISCUSSION

The analysis of polymeric and gaseous products of the reaction of PMMA and dimethylamine at low dimethylamine levels shows that there was a clean reaction that produced 1 eq of anhydride (NOT Amide) per 1 eq of dimethylamine and per 1 eq of MMA. Note that this is in most close agreement with pathway 2a→3a (Figure 2) in which most of the MMA carbonyls have been converted to COOH or alkylammonium COO^- salts followed by closure to anhydride. This shows that direct acyl addition (2c→3c) to form amide, or reaction of carboxylate with adjacent ester (2b→3b) to form anhydride (which liberates MeOH after proton transfer, and requires only 0.5 eq. dimethylamine per MMA group) are not the predominant pathways under these conditions. The Flory limit effect is consistent with random initial MAA formation, which after cyclization to anhydride (Figure 2) eventually leads to some groups trapped between rings which remain in the uncyclized acid form.

This confirms the previously proposed mechanism of the monomethylamine imidization reaction[7] except for the observed 1:1 dimethylamine/MMA stoichiometery. This stoichiometery indicates that the predominate reaction pathway of PMMA with dimethylamine probably involves two alkylations per anhydride (as in 2a → 3a, Figure 2). The formation of amide at extreme reaction conditions could be due to a slower competitive amine acyl addition pathway (2c or addition to acid). Other possibilities include equilibrium reactions where high amine pressures could shift the equilibrium from, for instance, anhydride (3a and 3b) to amic-acid pair (4a and 4b). If the latter concentration is high enough, or leads to some statistically trapped groups, then some amides might survive the devolatilization conditions.

Analysis of the gaseous products collected from sample 4 indicated trimethyl amine (93.9 meq) and H2O (95.5 meq) in a 1:1 ratio (86 % of products were from alkylation followed by elimination) and MeOH (15.3 meq, 14 % of products from ester condensation), confirming that alkyl-oxygen cleavage is indeed the predominant mechanism, and indicating that di-acid formation followed by elimination of water (2a → 3a) was about 6 times more likely than formation of the acid (or acid salt) followed by elimination of MeOH from adjacent ester (2b → 3b) or direct addition of dimethylamine to ester (2c → 3c). Also the low % N values at low dimethylamine ratios confirmed that little direct acyl addition was observed. Figure 3 shows a summary of the most probable reaction pathways. The low observed MeOH values in the dimethylamine case are in contrast to the previously studied imidization reaction with methylamine, where molar ratios of gaseous reaction products were 0.50/0.026/0.84/1.10-Trimethylamine/Dimethylamine/H2O/MeOH, indicating either predominant elimination of MeOH from esters adjacent to initially formed acid (salts) or, more likely, formation of the amic-acid pair with elimination of MeOH by a neighboring group effect in the case of the primary amine, but essentially NO diacid formation (Figure 4).

Although the first step of the mechanism (alkyl-oxygen cleavage of MMA) remains the same for mono- and di-methylamine, the shift in mechanism in the subsequent steps upon going from primary amine to secondary amine may be an indication of slower amic-acid formation due to the steric bulk of the dialkyl amine, or more rapid formation of trimethylammonium methacrylate (2.2), which converts most of the available adjacent ester groups to acid salts prior to the elimination stage. This could occur due to the much greater nucleophilicity of dimethylamine compared to monomethylamine. This would also account for the very small quantities of dimethylamine found in imidization reactions with monomethylamine. In that case, the dimethylamine formed from the first alkylation of monomethylamine rapidly alkylates again (forming trimethylamine), and is therefore found in only trace quantities. In the dimethylamine reactions studied herein, large quantities of dimethylamine are present which lead to many adjacent acid groups, which eliminate water leading to anhydride.

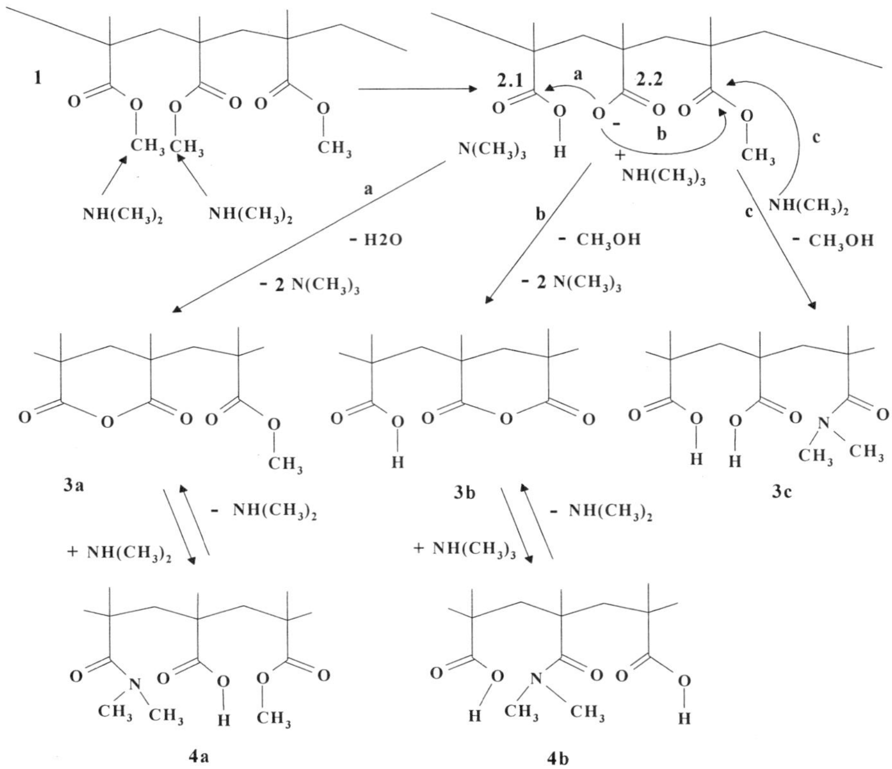

Figure 2. Possible Reaction Pathways of Dimethyamine and PMMA at 302°C

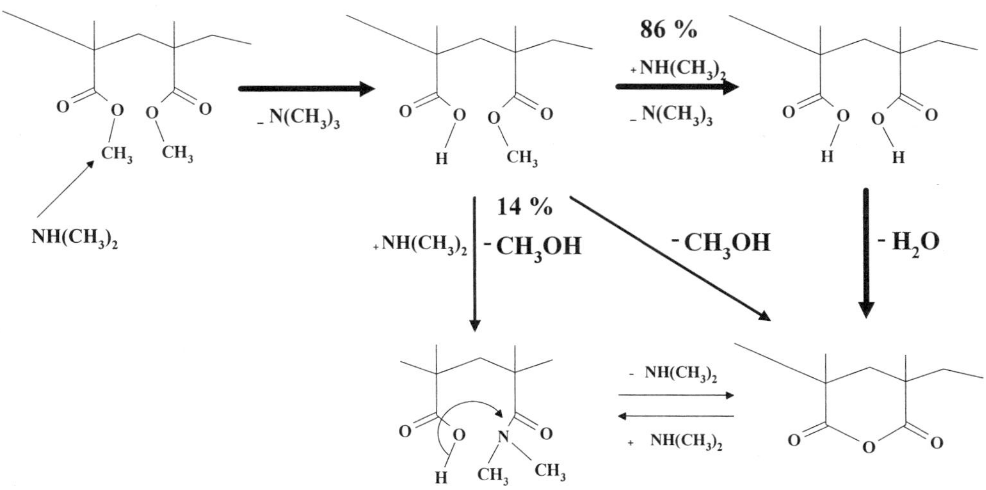

Figure 3. Most Probable Reaction Pathways of Dimethyamine and PMMA at 302°C

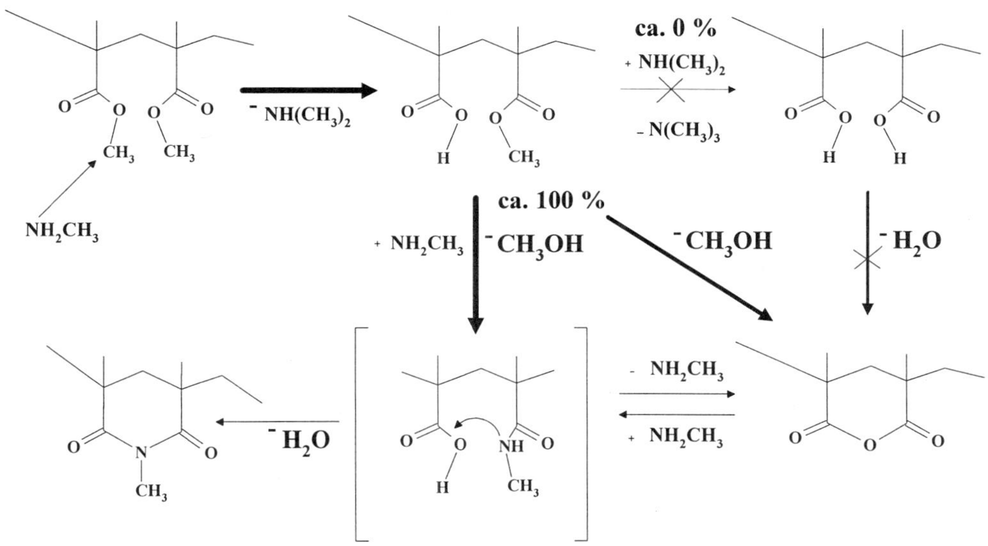

Figure 4. Most Probable Reaction Pathways of Monomethyamine and PMMA at 302°C

Figures 3 and 4 therefore represent our best understanding to date of the reaction pathway for PMMA and dimethyl- or monomethyl- amine respectively. Currently, molecular modeling work is underway (in conjunction with Dr. S. Fitzwater) to help understand these reactions in additional detail.

Acknowledgments

My thanks to R. Steffenino, E. White, M. Abramowicz, L. Cohen, R. Wood, W. Work, H. L. McPeters and S. Fitzwater for synthesis, analytical help, and general discussions.

REFERENCES

1. G. Graves, U.S. Pat. 2,146,209 (1939).
2. G. Schroder and K. Tessmar, U.S. Pat. 3,284,425 (1966).
3. F. Gotzen and G. Schroder, *Makromolekulare Chem*, **88:**133 (1965).
4. V.W. Ganzler, P. Huch, W. Metzger, and G. Schroder, *Angew. Makromolekulare Chem*, **11:**91 (1970).
5. R. M. Kopchik, U.S. Pat. 4,246,374 (1981).
6. M. Hallden-Abberton, N. Bortnick, L. Cohen, W. Freed, and H. Fromuth, U.S. Pat. 4,727,117 (1988).
7. M. Hallden-Abberton, *Polym. Mater. Sci. Eng.*, **65:**361 (1991).
8. H. Hiraoka, *Makromolecules*, **9:**359 (1976).
9. A. Jamieson and I. C. McNeill, *Eur. Poly. Journal*, **10:**217 (1973).
10. M. Hallden-Abberton, L. Cohen, R. Wood, U.S. Pat. 4,874,824.
11. E. Flexman, G.B. Pat. 2,101,139 (1982).

GRAFTING OF HINDERED AMINES GROUPS ON EPDM AND POLYOCTENAMER VIA PHOTO-HYDROPEROXIDATION

J. Lacoste[1], S. Chmela[2], J. Pellet[1] and J.F. Pilichowski[1]

[1]Laboratoire de Photochimie, URA CNRS 433,
Université B. Pascal et ENS Chimie de Clermont-Ferrand
63177 Aubière - France
[2]Polymer Institute, Slovak Academy of Sciences
842-36 Bratislava, Slovak Republik

ABSTRACT

This paper reports on the selective hydroperoxidation of unsaturated elastomers by using singlet oxygen photochemically produced (sensitization by anthracene at 365 nm). Hydroperoxide groups were characterized by using derivatization by nitric oxide and dimethyl sulfide. Grafting was performed by decomposing of hydroperoxide groups in the presence of acrylate monomers or nitroxyl radicals. According to the substituents, in acrylate hindered amine stabilizers (HAS) can be obtained. In the last part, the effect of grafted HAS on photo and thermal ageing of the parent elastomer is reported.

INTRODUCTION

The reaction of singlet oxygen with double bonds is a convenient way to produce hydroperoxides both from unsaturated low molecular weight compounds as well as unsaturated polymers[1-4]. The reaction is quite selective (low formation of carbonyl and alcoholic by-products) with a yield up to 20 and 40 % according to the substrate.

Singlet oxygen can be chemically or photochemically produced. In the latter case, specific photosensitizers and suitable wavelengths are required.

Hydroperoxides, can be afterwards easily decomposed by thermal or photo-irradiations in moderate conditions ($T \approx 80°C$ or $\lambda > 300$ nm)[5-8]. For example, these compounds are well known to be key products in autooxidation of hydrocarbons or in natural ageing of most polymers. Main reactions are as follows :

formation

$$P^{\bullet} + O_2 \longrightarrow POO^{\bullet}$$

$$POO^{\bullet} + PH \longrightarrow POOH + P^{\bullet} \;(\longrightarrow \ldots)$$

decomposition

$$POOH \xrightarrow{h\nu,\ \Delta} PO^{\bullet} + {}^{\bullet}OH$$

$$2POOH \longrightarrow PO^{\bullet} + POO^{\bullet} + H_2O$$

$\longrightarrow$ alcohols, ketones, acids.

When the decomposition is performed in the presence of unsaturated monomers (styrenic, acrylic...) a free-radical polymerisation can be initiated and grafting occurs :

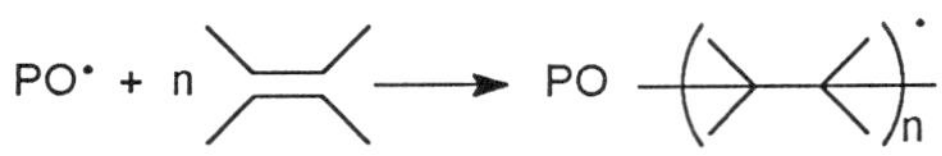

Moreover, when the decomposition is performed in the presence of nitroxyl compounds, the resulting radical/radical annihilation also involves grafting :

$$PO^{\bullet} + {}^{\bullet}ON\sim\sim \longrightarrow PON\sim\sim + 1/2\ O_2$$

If suitable substituents as sterically hindered piperidine groups are located on the grafted molecules, the synthesis of new polymeric stabilizers can be achieved. Such stabilizers can be expected to have a good efficiency in the protection of the parent polymer, especially in the case of thermal ageing.

In the present work, we performed the hydroperoxidation of unsaturated polymers as EPDM or polyoctenamer by using singlet oxygen photo-generated at 365 nm by anthracene / O_2. The resulting hydroperoxidized polymer (POOH) was then decomposed by UV irradiation ($\lambda > 300$ nm) in the presence of unsaturated monomers or nitroxyl radicals. The efficiency of these new grafted polymers as new photo- and thermo-stabilizers will be finally demonstrated.

EXPERIMENTAL

EPDM with different contents of 5-ethylidene-2-norbornene (VISTALON 7500 (VIST), EXXON, 0.7 mol. kg^{-1} ; POLYSAR 585, (POLY) HOECHST, 1.0 mol. kg^{-1}) and polyoctenamer (VESTENAMER, (VEST) HULS, 9.1 mol. kg^{-1}) were first precipitated from toluene solutions to remove possible additives.

Chemical modifications of the polymers were performed on films or in toluene solutions.

In the first case, thin films (*ca.* 0.1 - 0.15 mm) were prepared by compression moulding in an electrically heated laboratory press. Photo-hydroperoxidation were performed by : i) dipping the polymer film in a methanol solution of anthracene (*ca.* 3.10^{-3} mol. L^{-1}), ii) drying the polymer and irradiating the film in front of a source emitting at 365 ± 5 nm (MAZDA MAW, 125 watts). Photo-graftings were performed by irradiation of POOH film at $\lambda > 300$ nm, 35°C (MAZDA MA, 400 watts) in the presence of the vapour of volatile monomers* (methyl methacrylate, styrene, 1,2,2,6,6-pentamethyl-4-piperidyl acrylate (PPA)).

Another technique was to irradiate POOH films in methanol solutions of non-volatile

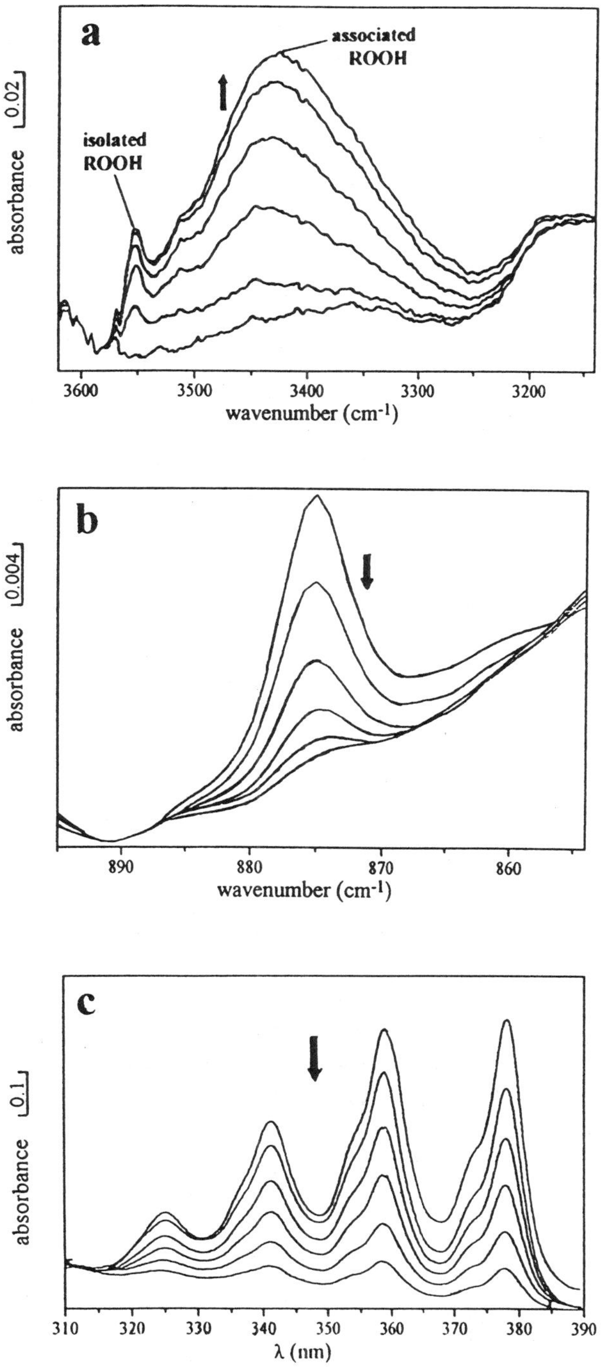

Figure 1. Spectroscopic changes upon hyproperoxidation of EPDM (POLY) films, exposure times each 20 min. (a) hydroperoxide grow ; (b) and (c) anthracene loss.

monomers* (*e.g.* 2,2,6,6-tetramethyl-4-piperidyl acrylate (TPA) and methacrylate (TPM)) or nitroxy radicals : 4-stearoyloxy-2,2,6,6-tetramethyl-piperidyl-N-oxy (St-NO•), 2,2,6,6-tetramethyl-piperidyl-N-oxy (TEMPO)** .

In the second case, anthracene was added to the toluene solution and irradiations were also performed at 365 nm. POOH can be precipitated into methanol. The grafting was made by irradiation at $\lambda > 300$ nm of the mixture POOH/monomer in toluene or *n*-heptane (typically 2 wt% solutions for each).

Anthracene consumption was followed by UV and IR spectroscopy (ε379 nm = 8900 L. $mol^{-1}cm^{-1}$). FTIR was used also to follow the formation of hydroperoxides and to detect grafting. Hydroperoxides were also characterized by derivatization by NO gas (converted into nitrates) or dimethyl sulfide (converted into alcohols, then nitrites) and by iodometric titration [12].

$$ROOH \xrightarrow{NO} RONO_2 \quad ; \quad ROOH \xrightarrow{DMS} ROH \xrightarrow{NO} RONO$$

For ageing experiments the toluene solution of the grafted polymer was first treated with an excess of triphenyl phosphine (reduction of residual hydroperoxides into alcohols), then suitable amount of the parent polymer was added and co-precipitation was performed into methanol. The polymer blend was vulcanized by adding dicumyl peroxide (5% w)prior precipitation (pressing conditions became 190°C, 190 bar, 5 min).
Another way was to introduce additives into a chloroform solution of the parent polymer, films being obtained by casting on glass plates[13]. Films were exposed to thermal- (oven at 80°C) and photoageing (irradiation chambers, described elsewhere[14], worked at $\lambda > 300$ nm and 35°C). Oxidations were followed by measuring changes in the hydroxyl and the carbonyl regions of IR spectra (NICOLET Impact 400, OMNIC Software).

RESULTS AND DISCUSSION

Polymer films

Fig 1 shows the evolution of the hydroperoxidation followed by FTIR and UV spectroscopy. Isolated, associated dimers and trimers and polyassociated hydroperoxides can be seen at 3550, 3520 (shoulder) and 3420 cm^{-1} respectively (fig. 1a).

Moreover, anthracene UV and IR bands decrease along the reaction because of an anthracene endoperoxide formation (Fig. 1b,c).

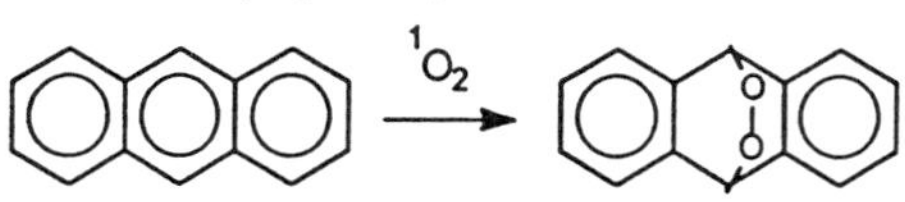

Hydroperoxide concentration was determined by iodometric titration, by FTIR (ε isolated ≈ εassociated ≈ 70 L $mol^{-1}cm^{-1}$) or by NO derivatisation (ε1276 cm₋₁ = 660 L $mol^{-1}cm^{-1}$) (see Fig. 2).

---

* MMA and styrene (PROLABO) were distilled prior grafting to remove inhibitors. The synthesis of piperidyl acrylates (TPA, PPA) and methacrylate (TPM) as well as the synthesis of St-NO• are described in ref. 9-11.

** TEMPO was obtained from ALDRICH.

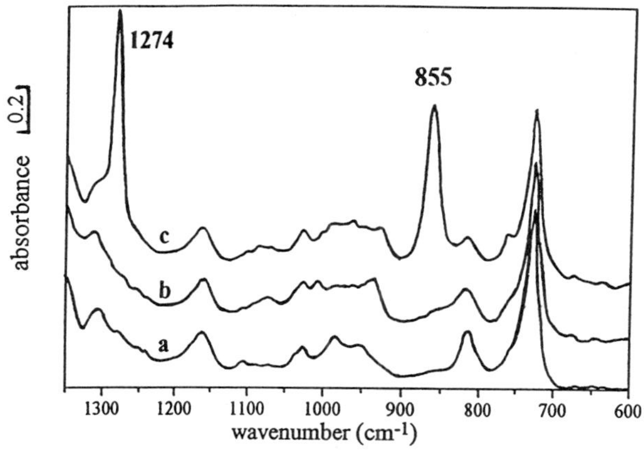

Figure 2. IR characterisation of EPDM (POLY)-OOH by NO derivatization. (a) control ; (b) EPDM-OOH ; (c) EPDM-ONO$_2$

The exact position of nitrate bands (at 1274 and 1276 cm^{-1} for EPDM and polyoctenamer respectively) is consistent with the selective formation of secondary hydroperoxides (a) [15].

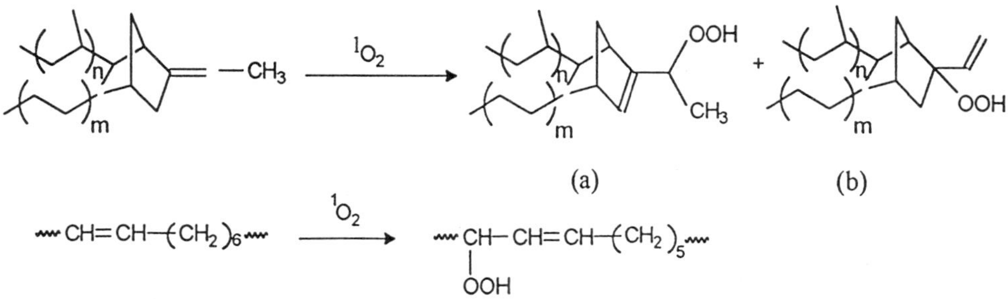

In the case of EPDM, an other possibility is the formation of a tertiary hydroperoxide (b) whose nitrate would be expected at 1300 cm^{-1} [15]. In addition, the absorption bands of ENB double bonds (extracyclic) are shifted significantly after hydroperoxidation as expected from the formation of an intracyclic double bond only. As it can be seen on fig. 3, the bands at 808 and 3040 cm^{-1} are shifted at 845 and 3082 cm^{-1} respectively. So secondary hydroperoxide (a) is the main product formed by singlet oxygen hydroperoxidation of EPDM.

Hydroperoxide concentrations such as 0.2, 0.35 and 0.85 mol kg^{-1} were measured ; that corresponds to a conversion of 25, 35 and 9 % of double bonds for VIST, POLY and VEST respectively. The hydroperoxide content can be increased by repeating the procedure of hydroperoxidation twice[12].

Photo-grafting of styrene, MMA and PPA were achieved by performing the photolysis of hydroperoxidized polymers in the presence of the monomer at low pressure (typically 5 mm Hg). Photo-grafting of PPA, TPA, TPM, St-NO• and TEMPO were performed in methanol solution of each monomers too. IR and NMR measurements ensured that the grafting was successful ; according the monomers, 1 to 6 units per one OOH group were grafteded.

Fig. 4 [10] shows FTIR spectra of some grafted EPDM films (each film was first washed with a suitable solvent to remove residual non-grafted homopolymers).

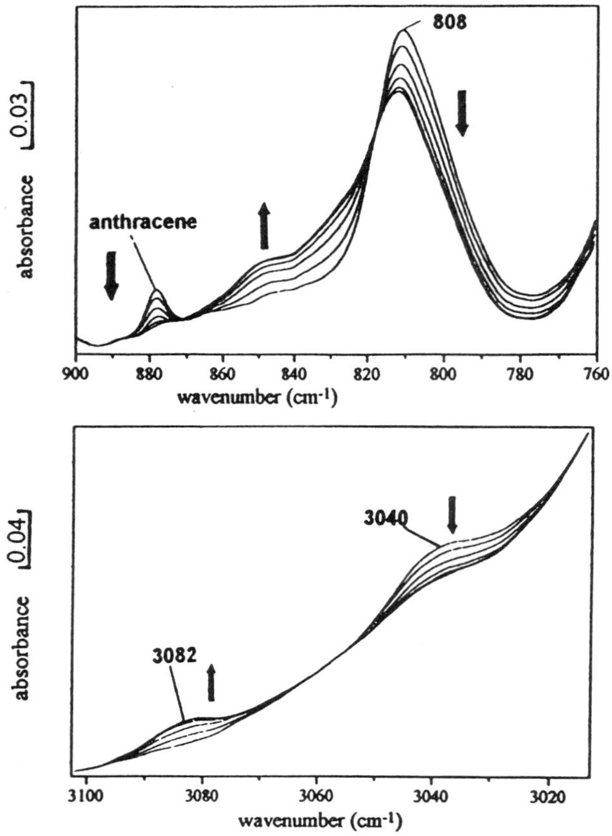

Figure 3. IR shifts upon hydroperoxidation of EPDM (POLY).

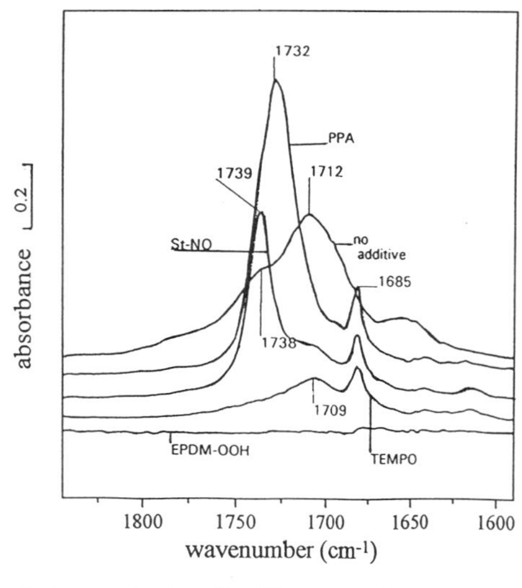

Figure 4. IR characterization of graftings on polyoctenamer (film thickness = 100 μm) (reproduced with permission from reference 10)

The homogeneity of both the hydroperoxidation and grafting was confirmed by FTIR microscopy across a cross section of the films[10].

Polymer solutions

Hydroperoxidation was also successful in the case of toluene solutions of both EPDM and polyoctenamer. Some differences in the course of the reaction were however revealed.

- the consumption of anthracene is much slower as in the case of the film and additional introduction of anthracene is not necessary to reach high hydroperoxide contents ;
- the reaction time is longer in toluene solution but much higher contents of hydroperoxide can be obtained ;
- the IR spectrum of the hydroxyl region naturally shows only a wide band located near 3500 cm^{-1} which belongs to the isolated hydroperoxides. Presence of isolated and associated hydroperoxides can be seen from IR spectra of polymer films after precipitation of the polymer into methanol (Fig. 5).

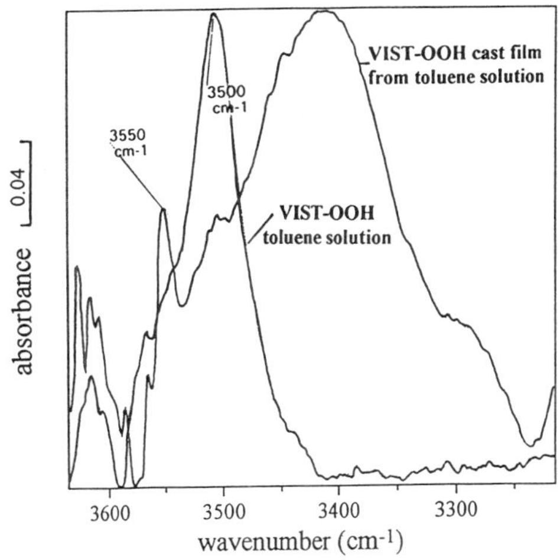

Figure 5. Comparison of IR spectra of EPDM (VIST)-OOH in film and in toluene solution.

The grafting in toluene solution is more efficient as in the case of film and crosslinking reactions (resulting also from the decomposition of hydroperoxide) are strongly reduced* . This last point is particularly important if we want to prepare polymeric additives able to be mixed properly with the parent polymer for ageing protection.

The grafting and the concomitent hydroperoxide loss can be followed by IR spectroscopy (ester band of acrylate (≈ 1735 cm^{-1}) and hydroperoxide band (≈ 3500 cm^{-1}) respectively), see Fig. 6.

The grafting efficiency of several elastomer and monomer pairs is reported on Table 1. In the case of POLY and VEST (containing high double bond contents) both hydroperoxidation and grafting were limited to avoid crosslinking (precipitation).

Ageing tests (from solution procedures)

Photoageing at λ>300 nm. The good photoprotective effect of polyoctenamer (VEST) grafted with PPA and St-NO• has been reported elsewhere[13].

* it was pointed out that, because of crosslinking, both hydroperoxidized and grafted films were becoming insoluble in toluene

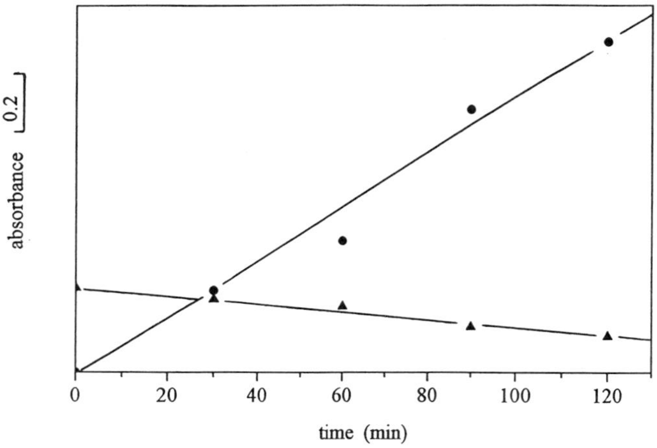

Figure 6. IR-evolution of ester (● 1680 cm^{-1}) and hydroperoxide (▲ 3400 cm^{-1}) upon grafting of TPA on EPDM (POLY).

Table 1. Grafting efficiency for films or solutions procedures.

	monomers units / ROOH*	
	films	toluene solutions
VIST-MMA	≈ 5	≈ 12
VIST-styrene	≈ 8	-
VIST-TPA	-	≈ 7
POLY-TPA	-	0.8**
VEST-PPA	-	≈ 3**
VEST-TPA	-	≈ 0.4**
VEST-St-NO$^{\bullet}$	-	≈ 0.4**

* initial content before reaction
**prior crosslinking

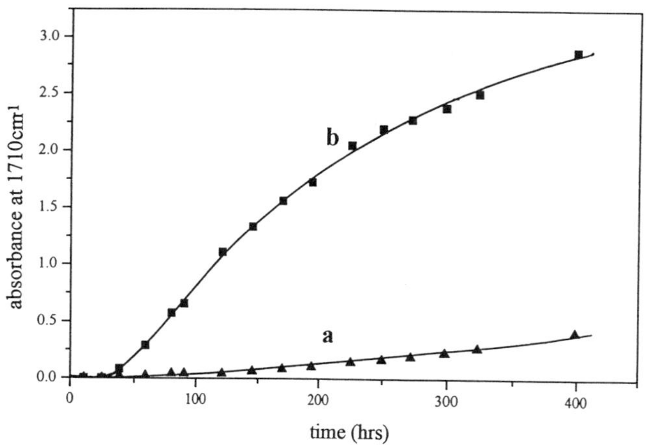

Figure 7. Efficiency of EPDM (POLY) grafted with TPA on the photoageing of EPDM (POLY). (a) control ; (b) containg 10% of EPDM-TPA.

The pronounced efficiency of EPDM (POLY) grafted with TPA on the photoageing of the parent polymer is reported on Figure 7. Similar experiment after vulcanisation (not shown) reduced the photo-oxidation rate by a factor 1.5 both for virgin and stabilized films.

Thermalageing at 80°C. The remarkable efficiency of EPDM (POLY) grafted with TPA can be seen on figure 8. Fig. 8a shows that the stabilization with 10 % w/w of EPDM-TPA in the crude elastomer. The effect of vulcanization is reported in fig. 8b, in this time scale the increase of stability is only seen on the virgin polymer.

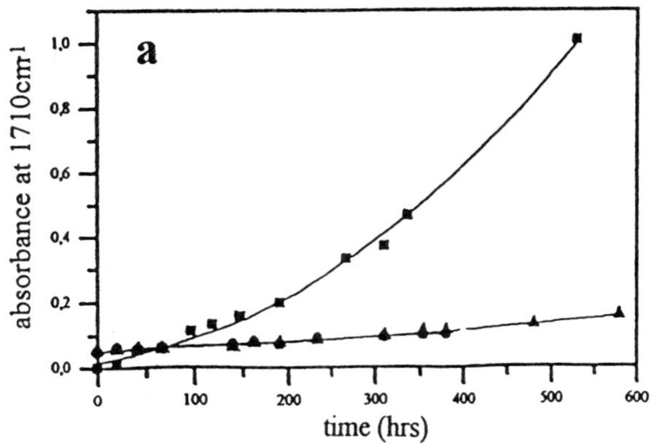

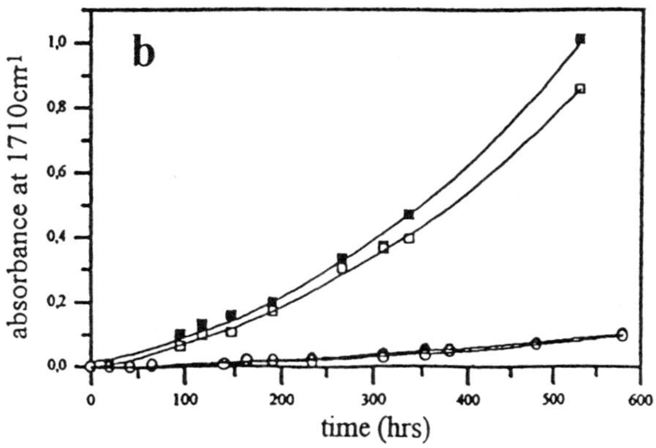

Figure 8. Efficiency of EPDM (POLY) grafted with TPA on the thermal ageing of EPDM (POLY). (a) effect of 10% w of EPDM-TPA ; (b) effect of additionnal vulcanization (empty squares and circles).

CONCLUSION

The grafting of unsaturated monomers or molecules bearing a nitroxyl radical group can be performed by using hydroperoxide as the intermediate. This group is sufficiently reactive to afford grafting by photochemical decomposition using mild conditions ($\lambda > 300$ nm, 35°C). Complementary experiments showed that thermografting is also efficient at temperatures above 80°C.

Given the importance of hydroperoxides in the oxidation (ageing) of most polymers, it is particularly interesting to use the instability of hydroperoxide to graft pendant groups as hindered amines whose stabilizing effect in polymer is well known. The concentration of grafted monomers can be regulated by the initial hydroperoxide concentration, that is, by the initial unsaturation index of the polymer or by the hydroperoxidation time. According to the decomposition reaction of hydroperoxide, three possible initiations can occur in the case of free radical polymerisation while in the grafting of nitroxyl radicals, only poly-alkyl substituted hydroxylamines are produced.

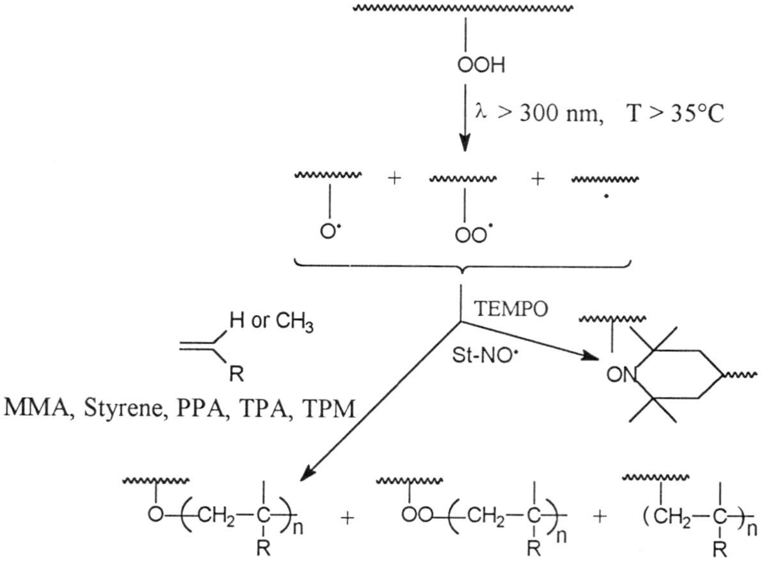

Finally, all the additives prepared by hydroperoxide intermediates increase dramatically the life time (both in photo and thermal ageing) of elastomers such as EPDM(ENB) and polyoctenamer whose the original stabilities are well known to be very low.

REFERENCES

1. C.S. Foote and G. Uhde, *Org. Photochem. Synt.* 1:60 (1971).
2. J.L. Courtneige, M. Bush and J. Loh, *J. Chem. Soc. Perkin Trans.* 1:1539 (1992).
3. R.W. Denny and A. Nickon, *Organic Reaction* 20:133 (1973).
4. G. Geuskens and M.N. Kanda, *Eur. Polym. J.* 27:877 (1991).
5. E.F.J. Duynstee and M.E. Mevis, *Eur. Polym. J.* 8:1375 (1972)
6. T.F. De Rosa, B.J. Kaufman and J.E. Broas, *J. Appl. Polym. Sci.* 42:1395 (1991).
7. D.J. Carlsson and D.M. Wiles, *Macromolecules* 2:597 (1969).
8. S. Commereuc, D. Vaillant, J.L. Philippart, J. Lacoste, J. Lemaire and D.J. Carlsson, *J. Polym. Deg. Stab.* in press
9. S. Chmela, P. Hardlovic and Z. Manasek, *J. Polym. Deg. Stab.* 11:233 (1985).
10. S. Chmela, J. Lacoste and G. Teissedre, *Macromolecules* 29:3055 (1996).
11. F.E. Karrer, *Makromol. Chem.* 181:595 (1980)
12. S. Chmela, G. Teissedre and J. Lacoste, *J. Polym. Sci., Part A, Polym. Chem.* 33:743 (1995).
13. S. Chmela, J.F. Pilichowski and J. Lacoste, *J. Polym. Sci. Polym. Chem. Ed.* in press
14. G. Teissedre, J.F. Pilichowski and J. Lacoste, *J. Polym. Deg. Stab.* 45:145 (1994).
15. D.J. Carlsson, R. Brousseau, Can Zhang and D.M. Wiles, *J. Polym. Deg. Stab.* 17:303 (1987).

REACTIVE GASES AS REAGENTS FOR POLYMER FILMS CHEMICAL MODIFICATIONS

J.F. Pilichowski[1], S. Commereuc[1], I. Lukac[2],
G. Teissedre[1] and J. Lacoste[1]

[1]Université B. Pascal et ENS Chimie de Clermont-Ferrand
63177 Aubière-France
[2]Polymer Institute, Slovak Academy of Science, CS-84236
Bratislava, Slovakia.

INTRODUCTION

Reactive gases as NO, SF_4, NH_3, SO_2 CH_2N_2... or the vapor of volatil liquids as CH_3COX, CH_3OH, CH_3SCH_3... have been extensively used for the identification and the titration of oxidized groups appearing during the ageing of polymers[1-12]. These reactions can be performed in the solid state (films) and at room temperature. Moreover, their implement is generally quite easy, it just consists to put gases in close contact with the polymer film for few hours. The exposure duration can be adjusted to control the progress of the reaction through the film thickness. Surface as well as "mass" modifications can be performed according the seeked application (membranes, painting, printing coatings...).

This work concerns mainly the modification of commercial polymers bearing hydroxy fonctions as alcohol, hydroperoxide or carboxylic acid, by reactive gases or liquid volatil compounds capable to penetrate in the polymer matrix. The modifications of membranes properties as gas permeability or surface tension will also be reported. Few examples will also concern the reaction of double bond with I_2 and HBr vapor as well as the oxidation of piperidine group by peracetic acid.

EXPERIMENTAL

Copolymers of ethylene (E) or propylene (P) with acrylic acid (AA) or methacrylic acid (MA) were obtained from SP2 (USA) and used as received (EAA, PAA, PEAA, PMA). Copolymers of ethylene and vinyl alcohol (EVOH) were prepared by hydrolysis (NaOH 0,5M in methanol, reflux, 48 h), of commercial ethylene-vinyl acetate copolymers (EVA, Elf-Atochem) containing 5, 9, 14, and 18 w% of acetate. Hydroperoxides of EPDM based on 5-ethylidene-2-norbornene (0.7 mol Kg^{-1}, Exxon) and polyoctenamer (Vestenamer, VEST, Hüls) resulted from a reaction of polymer films with singlet oxygen.

$^1O_2$ was generated by irradiation (UV light, 365 nm) of anthracene as photo-sensitizer according to published methods[11,12*]:

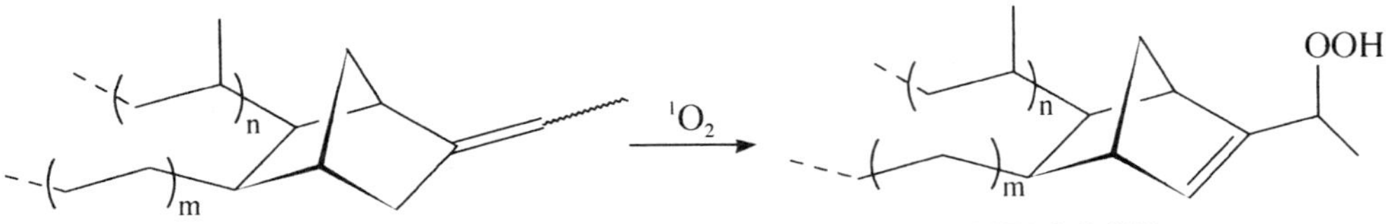

EPDM

EPDM-OOH

$$-[-CH=CH-(CH_2)_6-]_n- \xrightarrow{^1O_2} -[-\underset{}{\overset{OOH}{\overset{|}{C}}}H-CH=CH-(CH_2)_5-]_n-$$

VEST VEST-OOH

Gases [NO, NH_3 (Alphagaz), SF_4 (Fluka)], and volatil compounds [CH_3COCl, CH_3COF, CH_3OH, CH_3SCH_3, I_2, HBr and CH_3COOOH (Aldrich)], were used as received.

Films (50-100 μm) were obtained by compression molding in an electric heated laboratory press (θ = 90-180°C; P = 20-200 bars; t = 0.5-2.5 min).

Treatments by SF_4 were performed in a teflon flask equipped with Tygon inlet and outlet tubing. The gas was gently bubbling (1-2 min) and evacuated into a sodium hydroxide aqueous solution. The flask was then isolated with polyethylene stoppers and left to react 10-72 h. The flask was finally flushed (10-15 min) with dry nitrogen before FTIR analysis. Treatments by NO were performed in a borosilicate "trap" containing the polymer films. Air and moisture were first carefully removed by flushing (10-15 min) with dry nitrogen (NO is fast converted into NO_2 in the presence of oxygen). Then, NO was bubbling (1-2 min). The trap was finally isolated by stopcorks and left to react 3-4 days, in the dark, at room temperature. Ammonia treatment was indifferently realised with both apparatus.

Reaction with volatil liquids was performed in a small "laboratory dessicator" in which few drops of the liquid were deposited beside the polymer film. Derivatization reactions were followed by transmission and reflexion FTIR (Nicolet Impact 400 / Omnic software) or by ESR spectroscopy (Bruker ER200D)

In the particular case of the hydrobromination of VEST, the polymer was first dissolved into chloroform. Gazeous bromhydric acid was introduced at -18°C. Complete reaction (3-5 days at 8°C) resulted in the formation of brominated polyethylene :

$$\sim(CH_2)_6-CH=CH\sim \xrightarrow{HBr} \sim(CH_2)_7-\underset{Br}{\underset{|}{C}}H\sim$$

In the presence of 4-amino-2,2,6,6-tetramethyl-piperidine, a partial nucleophilic substitution of the bromine atom occured; it resulted in the formation of a new polymeric stabilizer which has been treated by peracetic acid vapor.

Gas permeabilities for O_2, N_2 and CO_2 were realised with Lyssy GPM 20 apparatus. Surface tensions of modified and unmodified polymers films were derived from contact angle measurements which were evaluated with six liquids (H_2O, $HCONH_2$, Hg, CH_2I_2, tricresyl-phosphate and 1-bromonaphtalene).

* See also another chapter from this book

RESULTS AND DISCUSSION

Polymers containing hydroxy groups

Nitric oxide and acetyl chloride treatments of polymer containing hydroxyl groups result in the formation of nitrite and ester derivatives respectively :

$$ROH \xrightarrow{NO} RONO \quad ; \quad ROH \xrightarrow{CH_3COCl} ROCOCH_3$$

Fig. 1 shows the increase of nitrite band near 780 cm^{-1} for various alcohol contents in ethylene-vinyl alcohol copolymers. The analysis of the hydroxyl region of the IR spectra (not shown) indicated that the reaction was not quantitative (residual OH band). The precise analysis of this band ($\varepsilon_{3400} \approx 70$ l / mol. cm) as well as the nitrite band ($\varepsilon_{780} \approx 639$ l / mol. cm) allows to evaluate the reaction yield considering the total film thickness (Transmission IR). The values decrease when the OH content increases (0.75; 0.62; 0.59 ans 0.59 for vinyl alcohol contents 2.6, 4.9, 7.7 and 10.1% respectively). Complementary analysis by reflexion IR (HATR) showed that the first 5-8 μm (Germanium crystal) were fully transformed while the analysis of the first 20-25 μm (Zinc Selenide crystal) revealed a decrease of the yield from 1 to 0.5 when the alcohol content was increasing. Then, this treatment can be helpfull for surface modification of membranes.

The changes in surface properties were analysed both by contact angles and gas permeabilities measurements. The contact angles values for a family of typical liquids are reported on Table 1. The resulting surface tensions, γ_s, for PE, EVOH and EVO-NO (34,

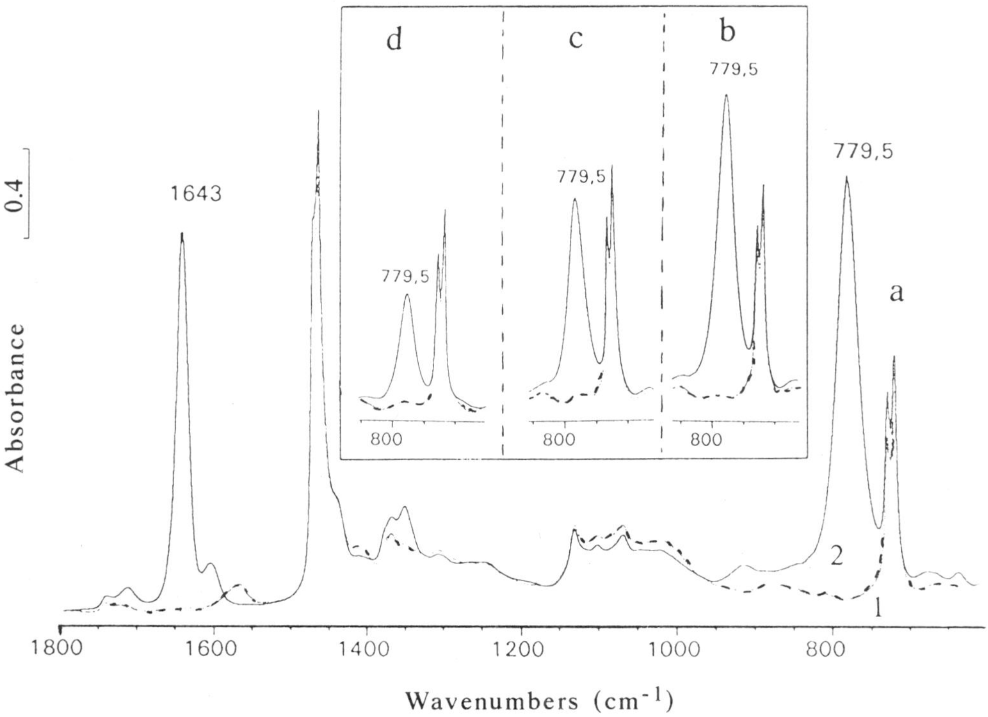

Figure 1. FTIR changes (1 ⇒ 2) upon NO treatment of ethylene-co-vinyl alcohol (EVOH) (a : 10.1; b : 7.7; c : 4.9; d : 2.6 w% vinyl alcool content).

39.5, and 43.2 mJ.m-2, respectively) can be calculated from a calibration curve ($\cos\theta = f(\gamma_l^{1/2}\gamma_l^{-1})$) based on the Fowkes relationship : $\cos\theta = 2(\gamma_s\gamma_l)^{1/2}\gamma_l^{-1} - 1$

Table 1. Contact angles values (deg.), of typical liquids with polymer surfaces.

Liquids	PE	EVOH (4.9 w%)	EVONO
Water	94	87	87
Formamide	77	63	77
Mercury	141	136	129
Di-iodomethane	43	41	43
Tri-cresylphosphate	45	32	30
1-Bromonaphtalene	22	24	19

Then, NO treatment results in a significant increase of the surface tension of ethylene-vinyl alcohol copolymers. This point was checked by gas permeability measurements. As shown on Table 2, both permeabilities and permeability coefficients (independent of thickness) increase significantly for "polar gases" (O_2 and CO_2).

Table 2. Permeability measurements; * values given by the instrument (in l / day / m^2 / atm) ** permeability coefficient (in 10^{-13}. cm^3 . cm / cm^2 . s. Pa).

	EVOH (10.1 w%)		EVO-NO	
	*	**	*	**
CO_2	4.2	6.2	8	11.8
O_2	1.4	2.1	1.9	2.8
N_2	1.2	1.8	0.6	0.89

Treatment by acetyl chloride is also very efficient. As shown on fig. 2, the reaction is practically complete after 24 h (disappearance of the OH band at 3375 cm^{-1} and appearance of the ester band at 1740 cm^{-1}) . For high hydroxyl contents, the conversion was much better than for nitric oxyde treatment probably because an higher reactivity for acetyl chloride.

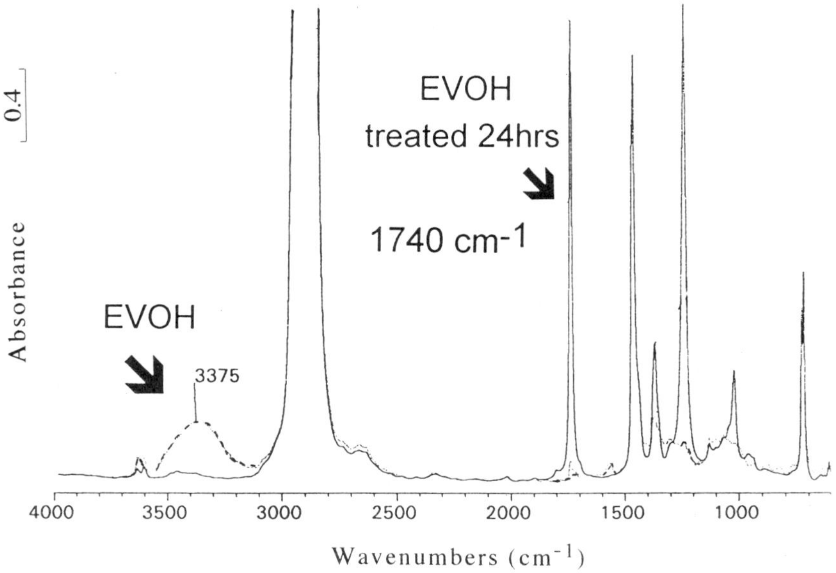

Figure 2. (re-) Generation of EVA from EVOH by acetyl chloride treatment.

Polymers containing hydroperoxide group

We studied dienic elastomer films (EPDM, poly-octenamer, poly-isobutylene-*co*-isoprene) which react with singlet oxygen, $^1O_2$, according a "ene" reaction described elsewhere[11-12]. The resulting hydroperoxidized polymers can also be treated by nitric oxide and acetyl chloride but the reactions result now in the formation of nitrate and perester respectively :

$$POOH \xrightarrow{NO} PONO_2 \quad ; \quad POOH \xrightarrow{CH_3COCl} POOCOCH_3$$

Results for EPDM are reported on fig. 3. A total conversion is obtained after 24 and 92 hrs for CH_3COCl and NO treatments respectively (complete disappearance of the OOH bands at 3550 and 3400 cm^{-1}). Typical nitrates (EPDM-ONO_2) and peresters (EPDM-$OOCOCH_3$) can be seen at 1276, 854... and 1752, 1244...cm^{-1} respectively (see fig. 3a). In the case of acetyl chloride treatment, a shoulder is also detected at 1740 cm^{-1} and may be attributed to the presence of some acetate groups. Permeability measurements for NO treated samples showed again a significant increase; values for O_2, CO_2 and N_2 evolve from 25, 7.8 and 3.5 (EPDM) to 62, 17 and 6.10^{-13}. cm^3. cm / cm^2. s. Pa (EPDM-ONO_2). This result is quite surprising if we consider the very low level in nitrate ($\approx$ 0.6 mol %).

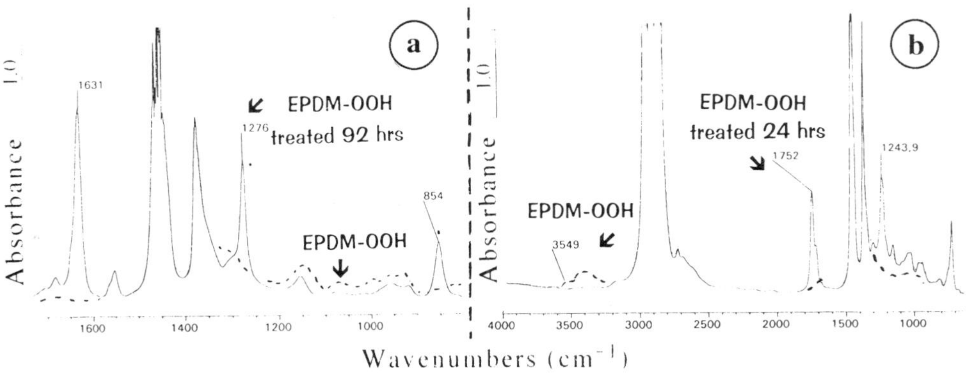

Figure 3. Transformation of EPDMOOH by a) NO and b) acetyl chloride treatments.

Moreover, by using dimethyl sulfide vapor, it is possible to reduce hydroperoxide groups into alcohols :

$$POOH \xrightarrow{DMS} POH$$

In this particular case, it is not easy to follow the reaction by FTIR spectroscopy as only a significant shift can be observed for weak bands of isolated groups (OOH_i : 3550 ; OH_i : 3620 cm^{-1}). However, an additional treatment by nitric oxide for both POOH and POH allows to follow the reaction since nitrates and nitrites have specific absorption (nitrates : 853, 1276 ; nitrites : 780 cm^{-1}) see fig. 5.

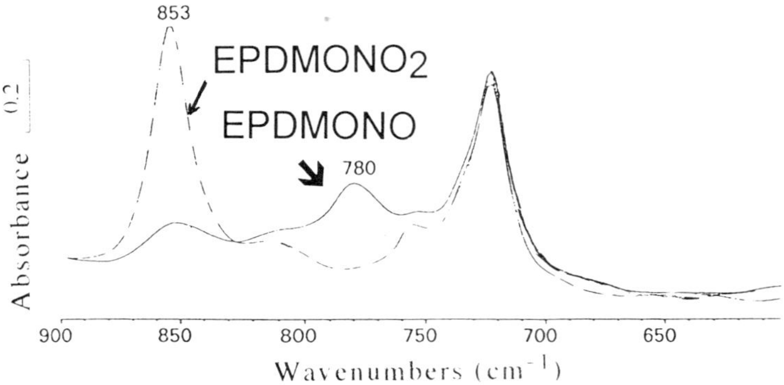

Figure 4. Main differences between hydroperoxidized EPDM after NO treatment without (EPDMONO2) and with preliminary reduction by gazeous DMS (EPDMONO).

Polymers containing carboxylic acid groups

Chemical modifications were also performed on copolymers of ethylene (E) or propylene (P) with acrylic (A) or methacrylic (M) acid. In a first time the gases used were sulfur tetrafluoride, acetyl chloride vapor and ammonia. The expected reaction are as follows :

$$P - COOH \begin{cases} \xrightarrow{SF_4} & P - COCl \\ \xrightarrow{CH_3COCl} & P - COOCH_3 \\ \xrightarrow{NH_3} & P - COO^- NH_4^+ \end{cases}$$

By adjusting the exposure time, quantitative conversion of carboxylic acid groups (followed near 1710 cm^{-1}) can be obtained (fig. 5). Typical bands at 1836, 1796 and 1555 cm^{-1} correspond to acid fluorides, acid chlorides and ammonium carboxylates groups respectively. In the case of EMA (Ethylene-Methacrylic acid) some sodium carboxylate groups were initially present. FTIR analysis showed that these functions (1559 cm^{-1}) were also converted into acid fluorides :

$$P{-}COO^- Na^+ \xrightarrow{SF4} P{-}COF$$

In the particular case of PAA (Propylene-acrylic acid) treated by SF4, an additional band has been observed at 1760 cm^{-1} which may be due to the presence of some anhydrides.

The conversion of carboxylic acids into corresponding acid halides (COF or COCl) involves an increase of the reactivity of organic acids and can authorize subsequent reactions not easy to perform in soft conditions with the parent acid. For example, amides and esters can be quantatively obtained by treatment of acid halide derivatives by ammonia and methanol vapor respectively :

$$P{-}COF \xrightarrow{NH3} P{-}CONH2 \; ; \; P{-}COCl \xrightarrow{CH3OH} P{-}COOCH3$$

Some examples are seen on fig. 6; amide and methyl ester groups are detected at 1639, 3210... and 1735 cm^{-1} respectively.

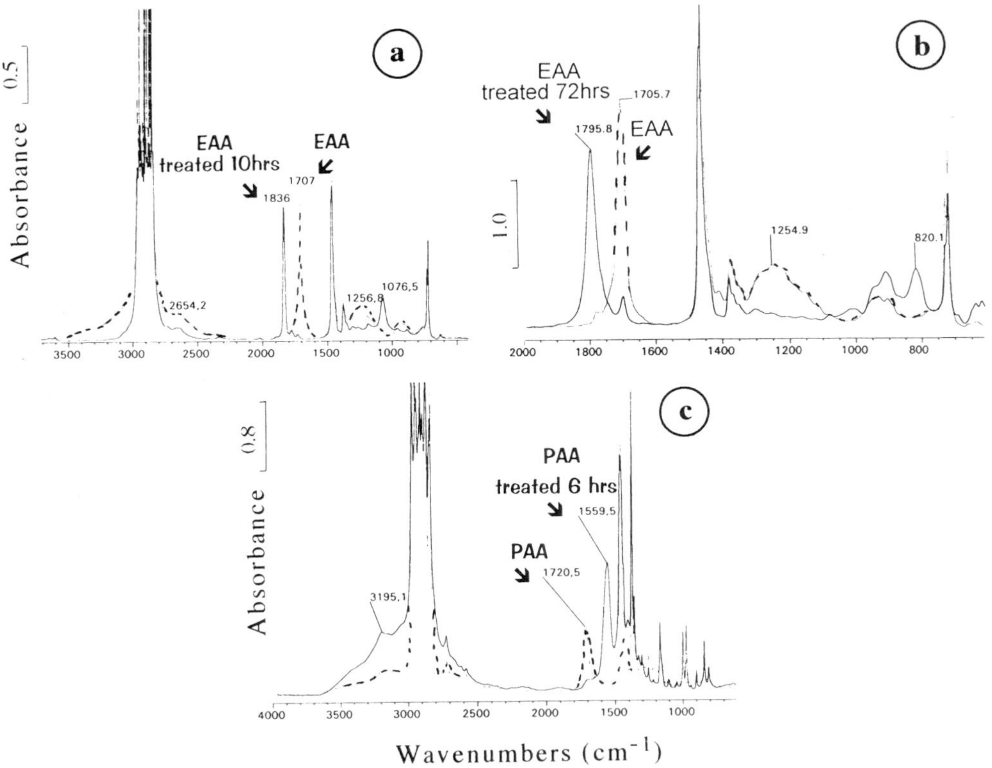

Figure 5. FTIR changes upon treatments of EAA by a) SF4; b) acetyl chloride and PAA c) by NH3.

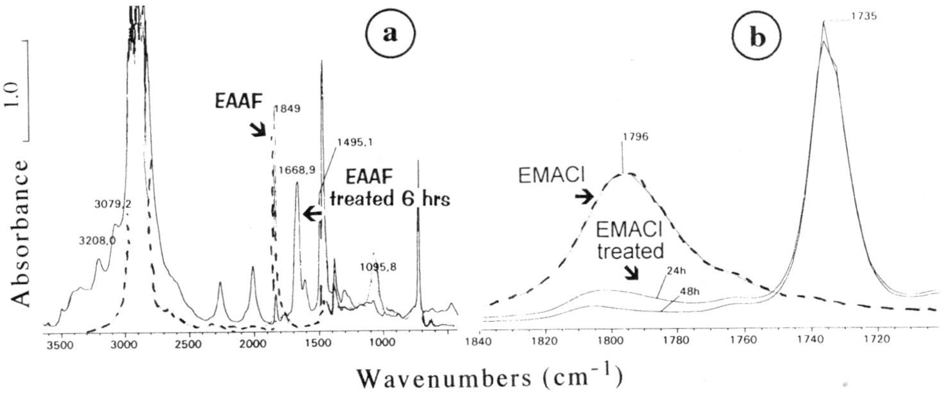

Figure 6. Transformations of acid halides : a) EAA-fluoride treated by MeOH ; b) EMA-chloride with NH3.

Polymers containing double bonds

The reactivity of a double bond, even in a polymer, is strong enough to authorize a lot of reactions (H_2, O_3...). We paid a special attention to the reaction of iodine vapor as it is a simple reaction for the IR identification of double bond bands in a polymer. As example, we can see on fig. 7 that bands associated with the double bond of ENB residue in EPDM can be identified at 3040, 1685, 808 and 540 cm^{-1}; the last two bands being assignable to E and Z forms respectively. The corresponding iodide derivative absorbs at 670 cm^{-1}.

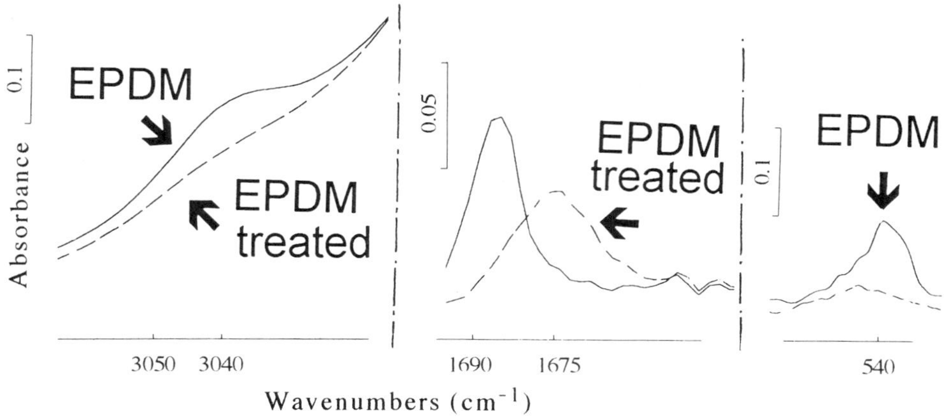

Figure 7. Identification of ENB double bond by treatment of EPDM with iodine vapor.

Moreover, the treatment of polyoctenamer film with hydrogen bromide gas was also tested. The yield was poor in these conditions; it is why the reaction was finally performed in a chloroform solution of the polymer (see exp. part).

Polymers containing sterically hindered piperidine groups

We recently synthesise a new polymer (see exp. part) containing significant amounts of tetra-methyl piperidine groups. Such a polymer is a very efficient polymeric thermal stabilizer for elastomer; however, when present in low content in the polymer matrix its IR detection is not easy because of the low absorption coefficients of all its functional groups.

An interesting alternative is to convert NH group into nitroxyl radicals by gaseous peracetic acid, these radicals being stable and easy to detect by ESR spectroscopy. Fig. 8 shows an example of the strong ESR signal obtained after a 24 h exposure.

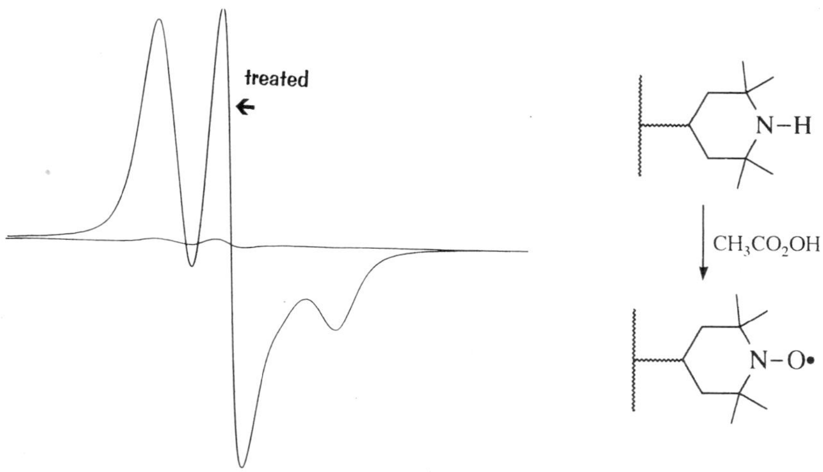

Figure 8. ESR signal observed during peracetic acid treatment of synthetic polymer bearing hindered piperidine group.

CONCLUSION

Treatment of polymer films by reactive gases or reactive volatil compounds allows to easily modify polymers containing alcohols, hydroperoxides, carboxylic acids (or acids halides), double bonds or piperidine groups. New functional groups as organic nitrites, nitrates, iodides, acid halides (Cl, F), amides, esters, peresters and nitroxyl radicals can be generated by a single reaction or by combination of two consecutive treatments. The reactions are very efficient on thin films (*ca* 50-100 μm) and can be controlled by transmission and reflexion IR spectroscopy.

Most of the functional groups obtained in this work have very high absorption coefficients comparatively with the parent functions and this technique can also be usefull to improve their detection in polymer matrix. From this point of view, reactions as saturation of double bonds or oxidation of NH group into nitroxyl radical can allow to follow these groups *in situ* along typical reactions as the ageing.

The following table summarize the reactions described :

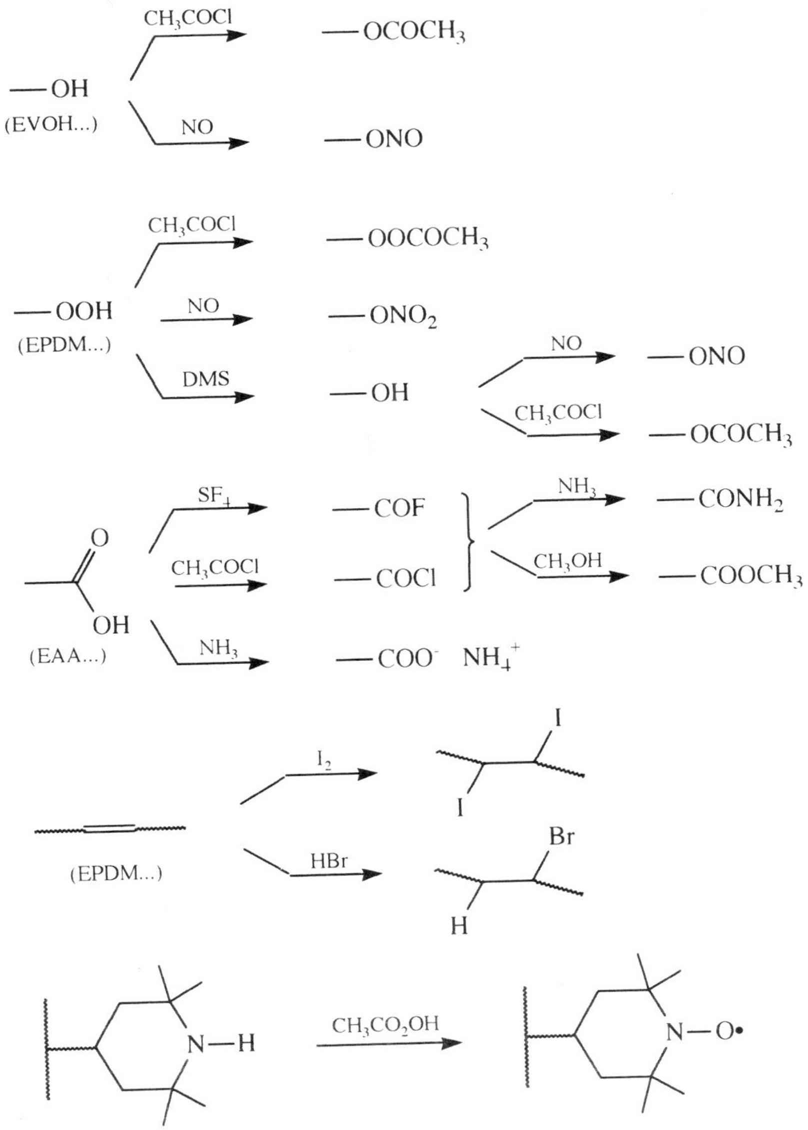

Acknowledgement

The authors express their thanks to ANVAR-FRANCE (Agence Nationale de Valorisation de la Recherche) for financial support.

REFERENCES

1. R.A. Nyquist and W.J. Potts, *Spectrochim. Acta* **17**: 679 (1961).
2. J.F. Heacock, *J. Polym. Sci.* **7**: 2319 (1963).
3. J.R. Rasmussen, E.R. Steoronsky and C.M. Whitesides, *J. Amer. Chem. Soc.* **99**: 4736 (1977).
4. J. Mitchell and L.R. Perkins, *Appl. Polym. Symp.* **4**: 167 (1967).
5. J.R. Shelton and R.F. Kopchzewski, *J. Org. Chem.* **32**: 2908 (1967).
6. L.J. Gerenser, J.F. Elman, M.G. Mason and J.M. Pochan, *Polymer* **26**:1162 (1985).
7. T.J. Henman, *Dev. Polym. Stab.* **6**: 107 (1995).
8. D.J. Carlsson, R. Brousseau, Can Zhang, and D.M. Wiles, *Polymer Deg. Stab.* **17**: 303 (1987) and *ACS Symposium Series* **364**: 376 (1988).
9. D.J. Carlsson, R. Brousseau and D.M. Wiles, *Polymer Deg. Stab.* **15**: 67 (1986).
10. Lacoste, D. Vaillant and D.J. Carlsson, *J. Polym. Sci. Part A: Polym. Chem* . **31**: 715 (1993).
11. S. Chmela, G. Teissedre and J. Lacoste, *J. Polym. Sci. Part A: Polym. Chem* . **33**: 743 (1995).
12. S. Chmela, J.F. Pilichowski and J. Lacoste, *Polym. Sci. Part A: Polym. Chem* . in press.

THE SYNTHESIS OF HYDROPHOBE-MODIFIED HYDROXYETHYL CELLULOSE POLYMERS USING PHASE TRANSFER CATALYSIS

Emmett M. Partain

Union Carbide Corporation
Bound Brook Technical Center
Bound Brook, NJ 08805

INTRODUCTION

Hydrophobe-modified water-soluble polymers have found extensive use in the latex paint industry as thickeners and rheology modifiers.[1] Hydrophobe-modified water-soluble polymers, also referred to as associative thickeners, consist of three general families of polymers: the HEUR (hydrophobe-modified ethoxylated urethane) thickeners, the HASE (hydrophobe-modified alkali-soluble emulsion) thickeners, and the HMHEC (hydrophobe-modified hydroxyethyl cellulose) thickeners.[2] Because of their ease of formulation, high thickening efficiency, good tint and colorant acceptability, and good general rheology, HMHEC polymers are widely used in North America in commercial latex paints, either alone or in combination with other thickeners, and often in exterior paint formulations.

The rheological properties of all HMHEC polymers are profoundly affected by the hydrophobe molar substitution (MS) and the hydrophobe chain length.[3] For any given hydrophobic moiety, there is a threshold hydrophobe MS below which there are no significant associative interactions. The most common phenomenological evidence for associative behavior is a dramatic increase in the solution viscosity of HMHEC polymers as a function of hydrophobe MS. The solution viscosity of HMHEC polymers continues to increase as a function of hydrophobe MS until the maximum limit of solubility is reached, as which point the HMHEC polymer becomes insoluble in water.[3a]

However, this threshold of associative behavior is also dependent on the carbon chain length of the hydrophobe moiety, with the onset of associative behavior occurring at lower hydrophobe MS values with longer carbon chain length hydrophobes. Further, if the carbon chain length is insufficiently long, then no significant associative behavior is observed at any MS value. Typically, the cut-off for associative behavior in hydrophobe-modified water-soluble polymers is an effective carbon chain length of about 12. For example, in Table 1, the 1% solution viscosities of HMHEC polymers prepared from various glycidyl ethers are shown. Even at relatively high hydrophobe MS values of 0.066 and 0.039, HEC modified with either phenyl glycidyl ether or *t*-butyl phenyl glycidyl ether affords no increase in viscosity compared to an unsubstituted HEC polymer. Thus these hydrophobes do not confer associative behavior on HMHEC polymers at these MS values.

Most cellulose ethers including HMHEC polymers are produced by heterogeneous slurry processes consisting of separate organic and aqueous/cellulose ether phases. Yokota of Daicel[4] studied the propoxylation of cellulose using an aqueous toluene/*t*-butyl alcohol diluent, and he reported a diluent phase separation in which essentially all the caustic partitioned into the lower aqueous phase during the heterogeneous reaction. After the addition of aqueous caustic, two phases form (three if one counts the solid cellulose as a separate phase), and essentially all the caustic, water, and cellulose are in the lower aqueous phase, while most of the organic components remain in the upper organic phase. This phase separation in the slurry synthesis of HMHEC is shown in Figure 4. Although not quantified, similar two-phase behavior has been noted in our synthesis of HMHEC in aqueous *t*-butyl alcohol.

All syntheses of commercial HMHEC polymers are examples of the Williamson ether synthesis in which a highly polar cellulose ether alkoxide adds to a hydrophobic (non-polar) electrophile in an S_N2 process. A basic principle from physical organic chemistry is that for two species to react, they must collide on a molecular level. In the case of HMHEC synthesis, this requisite collision is difficult to achieve: the electrophile (hydrophobe) is essentially only organic soluble and the nucleophile (cellulose ether alkoxide) is essentially only water soluble. Further, the nucleophile is a polymeric species, so polymer swelling behavior complicates the alkylation process. In classical organic synthesis, the problem of reacting different species of varying polarity and solubility would be solved by finding some mutual solvent to conduct the reaction under homogeneous conditions. However, because of the high viscosity of solutions of cellulose ethers, homogeneous reaction conditions are impractical for commercial production of HMHEC polymers.

Because the cellulose ether alkoxide is present entirely in the aqueous phase, the rate-limiting step may be the partitioning (phase transport) of the hydrophobic electrophile across the interface from the organic to aqueous phase. If the reaction rate is controlled by diffusion of the electrophile across the interface, then one would expect a correlation between water solubility of the hydrophobe and its alkylation efficiency. The fact that the actual alkylation reaction is probably occurring in the aqueous phase (or at the interface) yet the electrophile itself is principally soluble in the organic phase has important mechanistic ramifications. This type of synthetic problem, in which one reactant is water soluble and the other organic soluble, should be amenable to the techniques of phase transfer catalysis (PTC) to yield significant improvements in the alkylation efficiency.

The fundamental theory of phase transfer catalysis (PTC) has been reviewed extensively.[5] Rather than attempt to find a mutual solvent for all of the reactive species, an appropriate catalyst is identified which modifies the solubility characteristics of one of the reactive species relative to the phase in which it is poorly solubilized. The literature on the use of PTC in the preparation of nitriles, halides, ether, and dihalocarbenes is extensive. Although PTC in the synthesis of C- and O-alkylated organic compounds has been studied, the use of PTC in polymer synthesis or polymer modification is not as well studied. A general review of PTC in polymer synthesis was published by Mathias.[6] Fréchet[7] described the use of PTC in the modification of halogenated polymers such as poly(vinyl bromide), and Nishikubo and co-workers[8] disclosed the reaction of poly(chloromethylstyrene) with nucleophiles under PTC conditions. Liotta and co-workers[9] reported the O-alkylation of bituminous coal with either 1-bromoheptane or 1-bromooctadecane. Poor O-alkylation efficiencies were reported with alkali metal hydroxides but excellent reactivity and efficiencies were found with the use of quaternary ammonium hydroxides, especially tetrabutyl- and tetrahexylammonium hydroxides. These results are indeed noteworthy because coal is a mineral and is not thought of as a reactive and swellable polymer. Clearly if coal can be efficiently O-alkylated under PTC conditions, then efficient O-alkylation of cellulose ethers should also be possible.

Theory would predict that PTC should be useful in increasing the alkylation efficiency of hydrophobic electrophiles with cellulose ether alkoxides. However, there is very little previous work reported in using PTC in the preparation of cellulose ethers. Daly and co-workers[10] reported that quaternary ammonium salts were useful in catalyzing the heterogeneous benzylation of cellulose, but when we applied this technique to the DPGE alkylation of nascent HEC in aqueous *t*-butyl alcohol, the presence of catalytic amounts of tetramethylammonium chloride or tetrabutylammonium bromide actually afforded lower alkylation efficiencies.

This paper describes our efforts to apply phase transfer catalysis (PTC) to the synthesis of HMHEC polymers. The potential use of PTC in the manufacture of HMHEC polymers is important because a large portion of the manufacturing cost of HMHEC polymers is the cost of the hydrophobe reactant. Higher alkylation efficiencies would increase the overall reaction efficiency and thus reduce the overall cost of the final HMHEC product. Increased hydrophobe alkylation efficiencies would also reduce the volume of unreacted hydrophobe in the waste stream and reduce disposal costs.

EXPERIMENTAL

1-Octanol/Water Partitioning

A test tube (25 x 150 mm) is charged with 0.10 g (measured to the nearest ± 0.1 mg) of the alkyl phenyl glycidyl ether, 10.00 g of 1-octanol, and 10.00 g of distilled water. The mixture is thermally equilibrated at 25.0°C and then vigorously mixed. The phases are allowed to separate, and the lower aqueous phase is carefully removed. The aqueous phase (5.0 g, measured to the nearest ± 0.1 mg) is serially diluted to 10.00 ml with ethanol, and the alkyl phenyl glycidyl ether concentration is measured spectrophotometrically at 275 nm.

HMHEC Synthesis

The phase transfer catalyzed alkylation reaction of dodecyl phenyl glycidyl ether (DPGE) with hydroxyethyl cellulose (HEC) was studied as a mechanistic model for the general PTC reaction with cellulose ethers. In this way, the most effective phase transfer catalysts and optimum reaction concentrations could be identified. As a model cellulose ether, CELLOSIZE® HEC[11] was chosen, and the phase transfer catalysts chosen for evaluation were aqueous solutions of choline hydroxide, tetramethyl-, tetrabutyl-, tetrahexyl-, and benzyltrimethylammonium hydroxides. The molar A/HEC ratio (molar ratio of alkali to HEC) used was 0.50, the diluent to HEC (D/HEC) weight ratio was 7.4, and the reaction diluent was aqueous *t*-butyl alcohol. Because some of the quaternary ammonium hydroxide charges would be accompanied by large additions of water, the initial water content of the diluent was adjusted so that the final diluent composition would be about 14.4% water in *t*-butyl alcohol. The results of these experiments are summarized in Table 2.

HEC Reaction with DPGE. A 500 ml, three-necked round-bottomed flask was fitted with a stirring paddle and motor, a subsurface nitrogen inlet, a serum cap, and a Friedrich condenser with a mineral oil bubbler. The flask was charged with 29.55 g (27.0 g contained) of CELLOSIZE® hydroxyethyl cellulose (HEC) QP-3L, 173 g of *t*-butyl alcohol, and 16.5 g of distilled water. The slurry was stirred for one hour while purging with nitrogen to remove any entrained oxygen. The mixture was then causticized with 21.1 g of 40% aqueous benzyltrimethylammonium hydroxide, added dropwise with a syringe over 5 minutes. The mixture was stirred for one hour under nitrogen, and 3.40 g of dodecyl phenyl glycidyl ether (Shell Chemicals, purity of 84.5 weight %) was added by syringe over 5 minutes. The mixture was heated and stirred under nitrogen at reflux for 4.5 hours.

The mixture was then cooled to ambient temperature and neutralized by adding 5.0 g of glacial acetic acid by syringe. After stirring for 15 minutes, the polymer was collected by vacuum filtration through a fritted metal Buchner funnel. The polymer was washed in a Waring blender four times with 500 ml of 4:1 (by volume) acetone/water and twice with 500 ml of pure acetone. The polymer was dried overnight *in vacuo* at 50°C, affording 26.81 g of an off-white granular solid. The ash content was found to be 0.96% (as sodium acetate), the dodecyl phenoxy MS was found to be 0.013 by ultraviolet spectroscopy (extinction coefficient = 1526 M^{-1},cm^{-1}), and the 1% corrected viscosity was found to be 26 cP (Brookfield LVT, spindle #2, 30 rpm).

HEC Reaction with 1-Bromohexadecane. A 500 ml, three-necked round-bottomed flask was fitted with a stirring paddle and motor, a subsurface nitrogen inlet, a serum cap, and a Friedrich condenser with a mineral oil bubbler. The flask was charged with 29.55 g (27.0 g contained) of CELLOSIZE® hydroxyethyl cellulose (HEC) QP-300, 173 g of *t*-butyl alcohol, and 16.5 g of distilled water. The slurry was stirred for one hour while purging with nitrogen to remove any entrained oxygen. The mixture was then causticized with 21.1 g of 40% aqueous benzyltrimethylammonium hydroxide, added dropwise with a syringe over 5 minutes. The mixture was stirred for one hour under nitrogen, and 6.0 g of 1-bromohexadecane was added by syringe dropwise over 5 minutes. The mixture was heated and stirred under nitrogen at reflux for 4.5 hours.

The mixture was then cooled to ambient temperature and neutralized by adding 5.0 g of glacial acetic acid by syringe. After stirring for 15 minutes, the polymer was collected by vacuum filtration through a fritted metal Buchner funnel. The polymer was washed in a Waring blender four times with 500 ml of 4:1 (by volume) acetone/water and twice with 500 ml of pure acetone. The polymer was dried overnight *in vacuo* at 50°C, affording 25.38 g of an off-white granular solid. The ash content was found to be 5.73% (as sodium acetate), the hexadecyl content was found to be 0.98% by gas chromatography (equivalent MS of 0.012), and the 1% corrected viscosity was found to be 2160 cP (Brookfield LVT, spindle #3, 30 rpm).

DISCUSSION

The heterogeneous ethoxylation and alkylation of wood pulp to produce HMHEC is known to be a two-phase process in which the organic reactant (alkylating agent) is dissolved in the organic phase, while the cellulose alkoxide is suspended in the aqueous phase. Thus, given that the polarity of 1-octanol is representative of common organic diluents, the partitioning experiment of selected glycidyl ethers between water and 1-octanol should be a reasonable model to predict the distribution of these alkylation agents between the organic and aqueous phases in HMHEC synthesis.

To determine the alkylation efficiencies of these selected glycidyl ethers, syntheses of HMHEC polymers were conducted using wood pulp and ethylene oxide, using the same glycidyl ethers used in the partitioning experiments (Table 1). The details on the synthesis of HMHEC polymers directly from wood pulp and ethylene oxide have been reported elsewhere.[12]

Table 1. Comparison of the water solubility and alkylation efficiency of selected glycidyl ethers in HMHEC synthesis from wood pulp and ethylene oxide (EO MS of 3.5)

Glycidyl ether	Water solubility (weight %)	Hydrophobe MS	1% HMHEC viscosity	Alkylation efficiency
None	--	--	15 cP	--
Phenyl	1.63%	0.066	16 cP	44%
t-Butyl phenyl	0.835%	0.039	14 cP	26%
Nonyl phenyl	0.576%	0.022	279 cP	11%
Dodecyl phenyl	0.026%	0.018	2300 cP	8%
3-*n*-Pentadecenyl	0.116%	0.008	1300 cP	5%

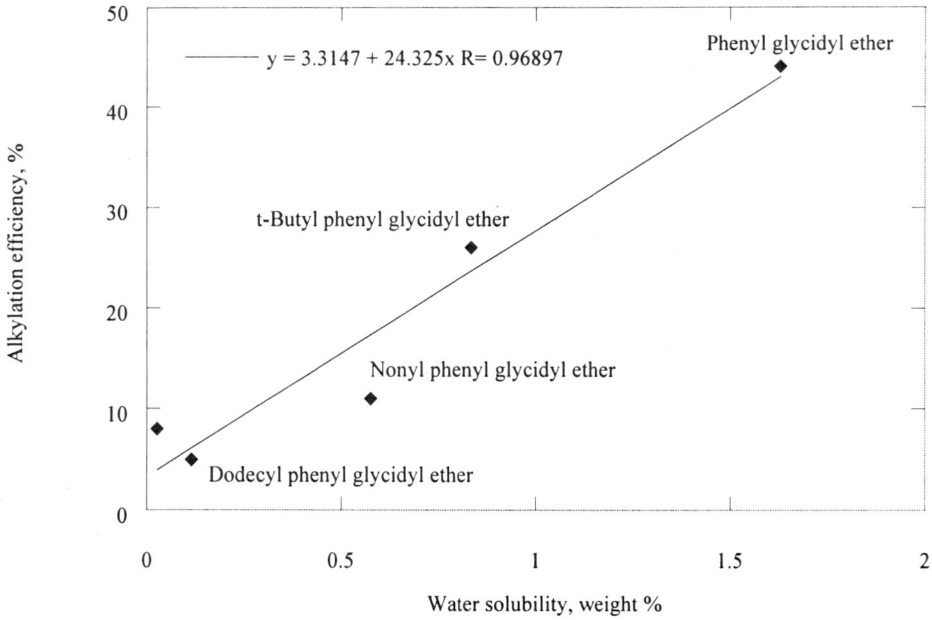

Figure 1. Effect of glycidyl ether water solubility on the alkylation efficiency in HMHEC synthesis from wood pulp and ethylene oxide

A plot of alkylation efficiency as a function of water solubility of the glycidyl ether (Figure 1) yields a linear relationship. This linear relationship, plus the apparent decrease in alkylation efficiency with increasing glycidyl ether carbon number, indicates that the alkylation efficiency of the glycidyl ether is dependent on its water solubility. Because the cellulose alkoxide is present only in the aqueous phase and because molecular collision is a prerequisite to a chemical reaction, these results imply that the rate-limiting step in the alkyl phenyl glycidyl ether alkylation of nascent HEC may be the diffusion of the alkylating agent across the interface from the organic phase to the aqueous phase.

Table 2. Synthesis of HMHEC polymers by phase transfer catalysis from CELLOSIZE® HEC QP-300 in aqueous *t*-butyl alcohol

Hydroxide	A/HEC molar ratio	Hydrophobe	Hydrophobe MS	Alkylation efficiency	1% Brookfield viscosity
Lithium	0.50	DPGE	0.007	5.9%	300 cP
Sodium	0.25	DPGE	0.005	4.1%	83 cP
Sodium	0.50	DPGE	0.008	6.8%	1188 cP
Sodium	0.75	DPGE	0.007	5.9%	577 cP
Potassium	0.50	DPGE	0.009	7.6%	1604 cP
Choline	0.50	DPGE	0.004	3.4%	62 cP
Tetramethylammonium	0.50	DPGE	0.011	9.3%	4360 cP
Tetrabutylammonium	0.25	DPGE	0.015	12.6%	> 7000 cP
Tetrabutylammonium	0.50	DPGE	0.014	11.8%	6340 cP
Tetrabutylammonium	0.75	DPGE	0.014	11.8%	4620 cP
Tetrahexylammonium	0.50	DPGE	0.012	10.1%	1904 cP
Benzyltrimethylammonium	0.15	DPGE	0.008	6.8%	319 cP
Benzyltrimethylammonium	0.25	DPGE	0.016	13.5%	4540 cP
Benzyltrimethylammonium	0.50	DPGE	0.016	13.5%	5500 cP
Benzyltrimethylammonium	0.75	DPGE	0.011	9.3%	2612 cP
Sodium	0.50	C_{16} Br	0.001	0.5%	50 cP
Benzyltrimethylammonium	0.50	C_{16} Br	0.012	6.2%	2160 cP
Sodium	0.50	PDPGE	0.002	1.9%	216 cP
Benzyltrimethylammonium	0.50	PDPGE	0.006	5.8%	1152 cP
Sodium	0.50	C_{16} α-OE	0.0001	0.063%	45 cP
Benzyltrimethylammonium	0.50	C_{16} α-OE	0.0002	0.13%	146 cP

DPGE = dodecyl phenyl glycidyl ether
C_{16} Br = 1-bromohexadecane
PDPGE = 3-*n*-pentadecenyl phenyl glycidyl ether
C_{16} α-OE = C_{16} alpha olefin epoxide

Baseline experiments with lithium, sodium, and potassium hydroxides afforded DPGE alkylation efficiencies of between about 5.9% and 7.6%. Choline hydroxide yielded a lower DPGE alkylation efficiency of 3.4%, while tetramethylammonium hydroxide yielded a modestly higher DPGE alkylation efficiency of 9.3%. However, tetrabutyl-, tetrahexyl-, and benzyltrimethylammonium hydroxides afforded significantly higher DPGE alkylation efficiencies of 11.8%, 10.1%, and 13.5%, respectively. Thus, these quaternary ammonium hydroxides afford about a two-fold increase in the DPGE alkylation efficiency (see Figure 2). It is not clear why the alkylation efficiency of DPGE decreases with higher carbon number quaternary ammonium hydroxides. Perhaps the enhanced organic solubility of tetrabutyl- and tetrahexylammonium hydroxides yields more undesired DPGE by-products, or these quaternary ammonium hydroxides may exhibit limited thermal stability because of possible Hoffman elimination (E_2, β elimination), although a E_i mechanism is also possible.[13] Such Hoffman-type β eliminations are not possible with benzyltrimethylammonium hydroxide.

The best alkylation efficiencies of DPGE were obtained with benzyltrimethyl- and tetrabutylammonium hydroxide. To explore the effect of the variation of A/HEC ratio on DPGE alkylation efficiency, experiments were conducted at varying A/HEC ratios (see Figure 3). The alkylation efficiency maximum occurs between about 0.25 and 0.50 A/HEC molar ratio. The observed alkylation efficiencies of DPGE with tetrabutylammonium hydroxide were comparable to the alkylation efficiencies with benzyltrimethylammonium hydroxide.

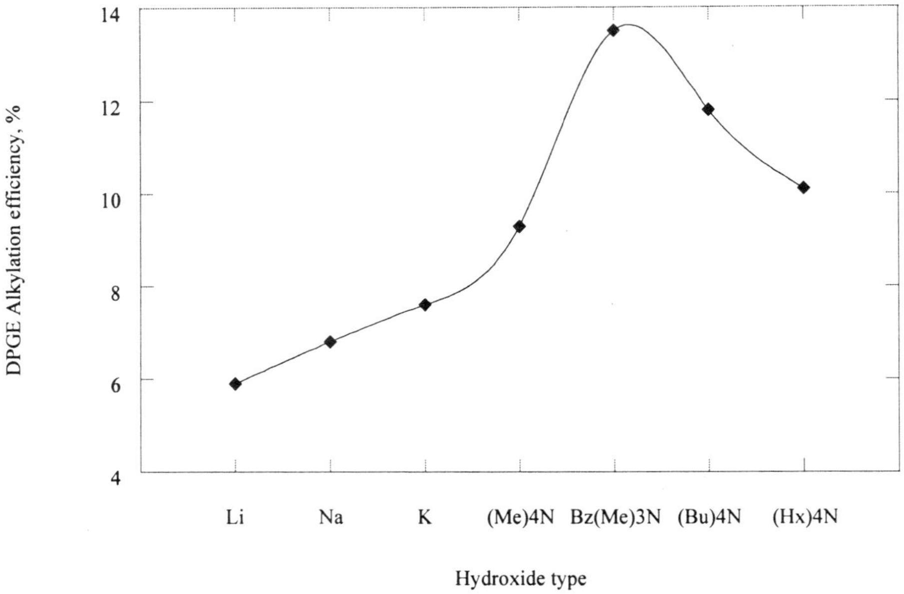

Figure 2. Effect of various hydroxide catalysts on the DPGE alkylation efficiency with HEC in aqueous *t*-butyl alcohol (A/HEC = 0.50)

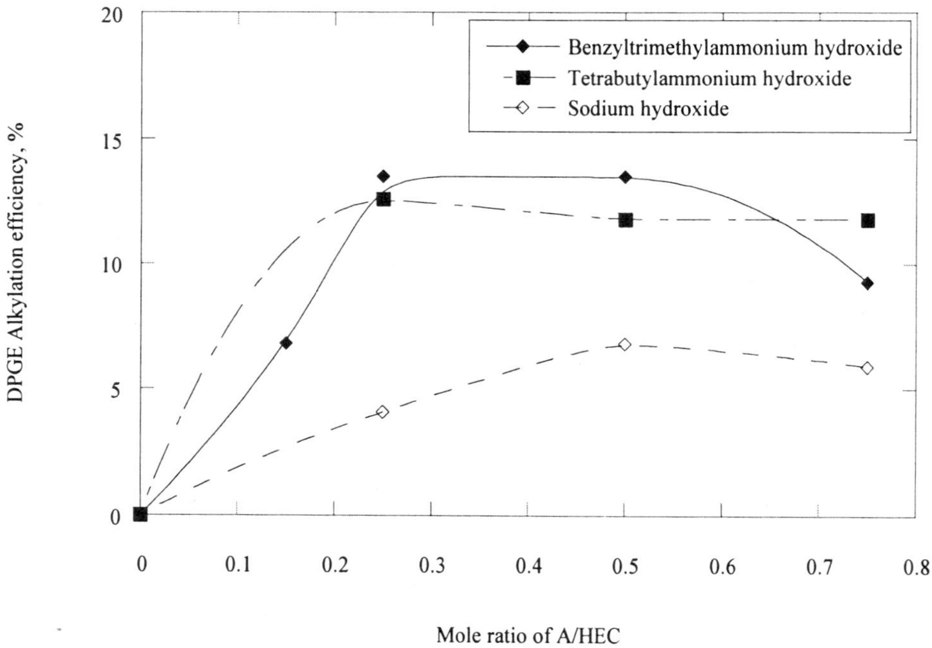

Figure 3. Effect of molar A/HEC ratio on the DPGE alkylation efficiency with HEC in aqueous *t*-butyl alcohol

The increase in alkylation efficiency of HEC with quaternary ammonium hydroxide is not limited to DPGE. Experiments were conducted in which HEC was alkylated with 1-bromohexadecane (cetyl bromide), 3-*n*-pentadecenyl phenyl glycidyl ether (PDPGE)[14], or 1,2-epoxyhexadecane (C_{16} α-olefin epoxide) in the presence of either sodium hydroxide or benzyltrimethylammonium hydroxide. As before, the molar A/HEC ratio was 0.50, and the water content of the diluent in the benzyltrimethylammonium hydroxide experiment was decreased to compensate for the higher water content of the base so that the final water content of both reactions was the same (14.4%). The hexadecyl content of the polymers was measured by gas chromatography. The sodium hydroxide mediated reaction of 1-bromohexadecane yielded a hexadecyl alkylation efficiency of 0.5%, while the benzyltrimethylammonium hydroxide reaction yielded a hexadecyl alkylation efficiency of 6.2%. A twelve-fold increase in the hexadecyl alkylation efficiency was observed in the reaction conducted with the quaternary ammonium hydroxide.

For PDPGE, the sodium hydroxide mediated alkylation reaction afforded an efficiency of 1.9% versus 5.8% for the benzyltrimethylammonium hydroxide reaction. Although the use of benzyltrimethylammonium hydroxide afforded a two-fold increase in the alkylation efficiency of HEC with 1,2-epoxyhexadecane, the efficiencies in either case were poor (0.063% versus 0.13%). However, 1,2-epoxyhexadecane is not very reactive to nucleophilic attack by alkoxide. In general, one would expect aryl glycidyl ethers to be more reactive to alkoxide nucleophilic attack than alkyl glycidyl ethers, which are in turn more reactive than alkyl epoxides.[15] A comparison of the alkylation efficiencies of various hydrophobic alkylating agents with HEC in aqueous *t*-butyl alcohol at an A/HEC ratio of 0.50 can be found in Table 3. Thus the improvement in the alkylation of HEC with benzyltrimethylammonium hydroxide as a representative quaternary ammonium hydroxide has been demonstrated with three different types of alkylating agents: an alkyl epoxide, two alkyl aryl glycidyl ethers, and an alkyl halide.

Table 3. Comparison of the sodium hydroxide and benzyltrimethylammonium hydroxide catalyzed alkylation of HEC in aqueous *t*-butyl alcohol

Alkylating agent	Alkylation efficiency	
	Sodium hydroxide	Benzyltrimethylammonium hydroxide
DPGE	6.8%	13.5%
PDPGE	1.9%	5.8%
1-Bromohexadecane	0.5%	6.2%
1,2-Epoxyhexadecane	0.063%	0.13%

The increase in alkylation efficiency noted with quaternary ammonium hydroxides in aqueous *t*-butyl alcohol was also observed in an aqueous acetone diluent. Experiments were conducted in which HEC was alkylated with DPGE in aqueous acetone diluent (85°C for three hours) using both 50% sodium hydroxide and 40% benzyltrimethylammonium hydroxide. As before, the molar A/HEC ratio was 0.50, and the water content of the diluent in the benzyltrimethylammonium hydroxide experiment was decreased to compensate for the higher water content of the base so that the final water content of both reactions was the same (11.4%). Sodium hydroxide afforded a DPGE alkylation efficiency of 2.5%, versus 6.7% with benzyltrimethylammonium hydroxide, again showing a 2.7 fold increase in DPGE alkylation efficiency with the quaternary ammonium hydroxide base.

The proposed overall mechanism is shown in Figure 4. The initial step in the alkylation reaction of a cellulose ether (or cellulose itself) is the conversion of the cellulose ether to the cellulose ether alkoxide. This is an equilibrium reaction (constant of κ_1), but the equilibrium lies toward the alkoxide.[16] In the case of the sodium hydroxide mediated alkylation reaction, the resulting sodium cellulose ether alkoxide is too polar to be appreciably soluble in the organic phase, so the only alkylation reaction that can occur is in the aqueous phase. The equilibrium partitioning of the alkylating agent from the organic phase to the aqueous phase (constant of κ_2) becomes the rate-limiting step, with final alkylated product being formed in the aqueous phase (rate constant of κ_6) by an S_N2 type reaction.

In the quaternary ammonium hydroxide mediated alkylation case, it is not clear what fraction of the alkylation reaction occurs in the organic phase versus the aqueous phase, and whether the majority of the reaction occurs in the organic phase itself or at the

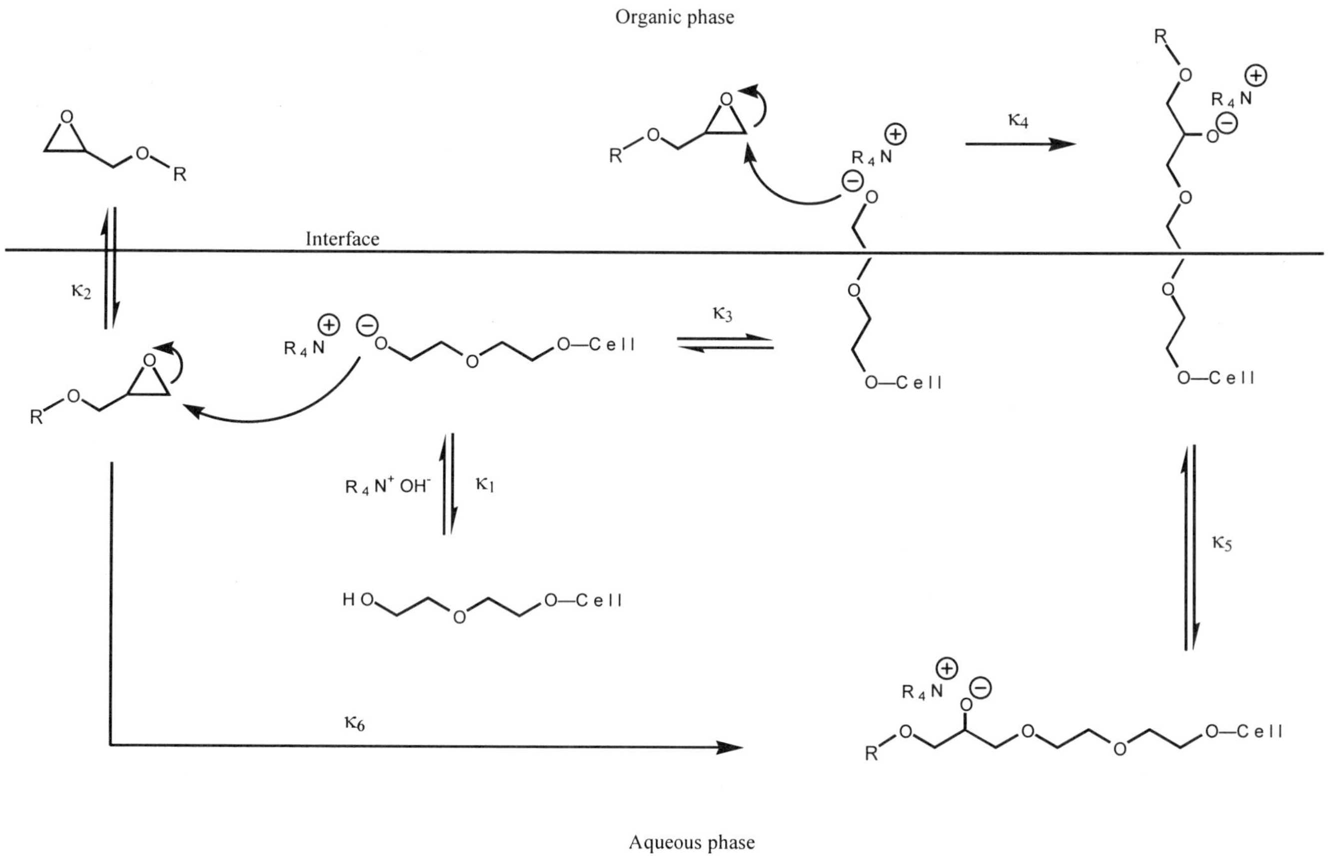

Figure 4. Phase transfer catalyzed hydrophobe alkylation of hydroxyethyl cellulose

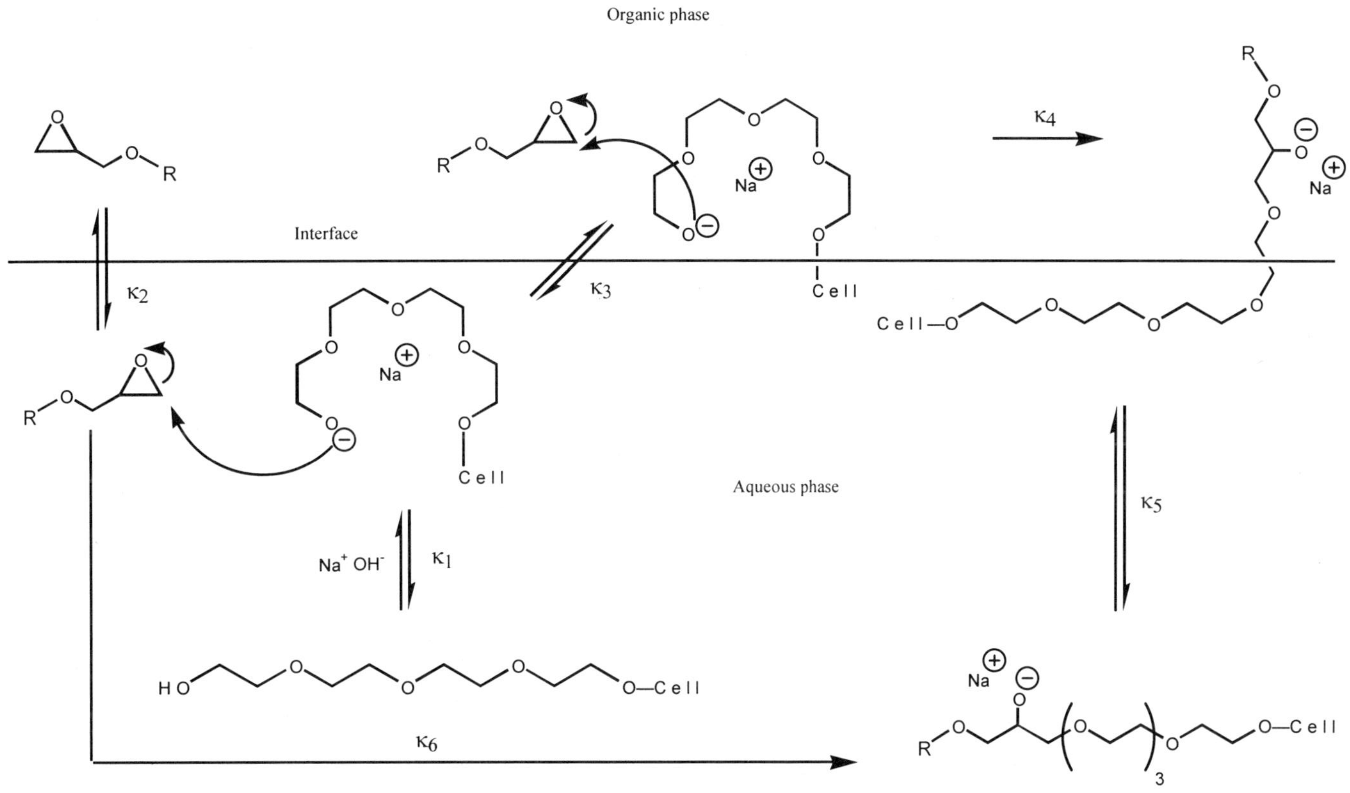

Figure 5. Pseudo-crown ether phase transfer mechanism

aqueous/organic interface. In either case, the quaternary ammonium cellulose ether alkoxide does exhibit some organic solubility, so a separate aqueous/organic (or aqueous/interfacial) partitioning equilibrium (constant of κ_3) is also in effect. Some fraction of the quaternary ammonium cellulose ether alkoxide in the organic phase or interface can react with the alkylating agent by an S_N2 process (rate constant κ_4), and final alkylated cellulose ether can then transfer back across the interface into the aqueous phase (constant κ_5).

The increase in DPGE alkylation efficiency with the use of quaternary ammonium hydroxides is clearly indicative of a PTC mechanism, although previously published work on HMHEC polymer synthesis also points to this result.[17] It has been shown that, in the absence of a formal phase transfer catalyst, the alkylation efficiency of hydrophobic electrophiles is controlled primarily by the EO MS of the HEC substrate. The preparation of HEC at higher EO MS values would tend to favor catenation of the EO chains and give longer EO chains. Initially, it was thought that the increase in alkylation efficiency was due to a reduction in steric hindrance and increase in lipophilicity of the pendent EO chains. However, in view of this PTC mechanism, the pendent EO chain could be considered a podand, an open-ended crown ether. This pseudo-crown ether phase transfer mechanism is shown in Figure 5. In this case, the long pendent EO chain, acting as a pseudo-crown ether would function as a built-in phase transfer catalyst.

Rather than using quaternary ammonium hydroxides directly in DPGE alkylation of HEC, experiments were conducted to generate these materials *in situ* from equimolar charges of sodium hydroxide and suitable quaternary ammonium halides. However, when this was tried with tetrabutylammonium bromide or benzyltrimethylammonium chloride (see Table 4), the DPGE alkylation efficiency was either equivalent to or less than the expected alkylation efficiency with sodium hydroxide alone. No phase transfer enhancement of the DPGE alkylation was observed. This phenomenon was attributed to solubility of the quaternary ammonium halide in the organic phase. In the aqueous phase, both sodium hydroxide and the quaternary ammonium hydroxides should ionize to a high degree. However, where both sodium hydroxide and the corresponding sodium halide have limited organic solubility, these quaternary ammonium halides exhibit substantial organic solubility. Thus, the added quaternary ammonium halides may be substantially extracted into the organic phase, so little quaternary ammonium halide remains in the organic phase to form the desired quaternary ammonium hydroxide. The high solubility of these quaternary ammonium halides in aqueous *t*-butyl alcohol may also explain the failure of tetramethylammonium chloride or tetrabutylammonium bromide to catalyze the DPGE alkylation of HEC.

Table 4. DPGE alkylation of HEC in aqueous *t*-butyl alcohol using quaternary ammonium salts (A/HEC ratio = 0.50)

Hydroxide	Added catalyst	Hydrophobe MS	Alkylation efficiency	1% Brookfield viscosity
Sodium	None	0.008	6.8%	1188 cP
Sodium	$Bz(Me)_3N^+ Cl^-$	0.008	6.8%	625 cP
Sodium	$(Bu)_4N^+ Br^-$	0.005	4.2%	100 cP
Benzyltrimethylammonium	None	0.016	13.5%	5500 cP
Benzyltrimethylammonium	NaOAc	0.007	5.9%	203 cP
Benzyltrimethylammonium	Na_2SO_4	0.016	13.5%	4700 cP

The solubility of quaternary ammonium salts in the organic phase can profoundly affect the course of any phase transfer alkylation reaction, and the organic solubility of these quaternary ammonium salts is strongly dependent on the counterion. The estimated order of organic solubility of quaternary ammonium salts is[5a]:

$$ClO_4^- >> I^- > NO_3^- > Br^- > Cl^- > OAc^- > F^- \;\&\; OH^- > SO_4^{-2} > CO_3^{-2} > PO_4^{-3}$$

From this ranking, quaternary ammonium acetates, chlorides, nitrates, and bromides are apparently more soluble in organic solvents such as aqueous *t*-butyl alcohol than are the corresponding quaternary ammonium hydroxides. Thus these quaternary ammonium salts would be readily extracted into the organic phase. In addition, a high concentration of sodium acetate, chloride, bromide, or nitrate in the aqueous phase of a cellulose etherification reaction could potentially interfere with the phase transfer catalytic effect of quaternary ammonium hydroxides. When a quaternary ammonium hydroxide catalyst and a high concentration of one of these sodium salts are mixed in the aqueous phase, the quaternary ammonium salt formed in equilibrium will also exhibit higher organic solubility than the corresponding quaternary ammonium hydroxide and will be preferentially extracted into the organic phase. The effective concentration of the quaternary ammonium hydroxide in the aqueous/cellulose ether phase will be correspondingly reduced.

To confirm the deleterious effect of certain counterions on the phase transfer catalyzed DGPE alkylation of HEC, two experiments were conducted in which HEC was causticized with benzyltrimethylammonium hydroxide and reacted with DPGE in the presence of an equal molar amount of sodium acetate or sodium sulfate (an organic soluble and organic insoluble counterion, see Table 4). As predicted by theory, the reaction mixture containing sodium acetate afforded a lower DPGE alkylation efficiency (5.9%) compared to either a control or the reaction containing sodium sulfate (both 13.5%). While any benzyltrimethylammonium acetate formed in the equilibrium could be extracted into the organic phase, the corresponding benzyltrimethylammonium sulfate has almost no organic solubility and thus remains in the aqueous phase. These results are consistent with the mechanism proposed in Figure 6.

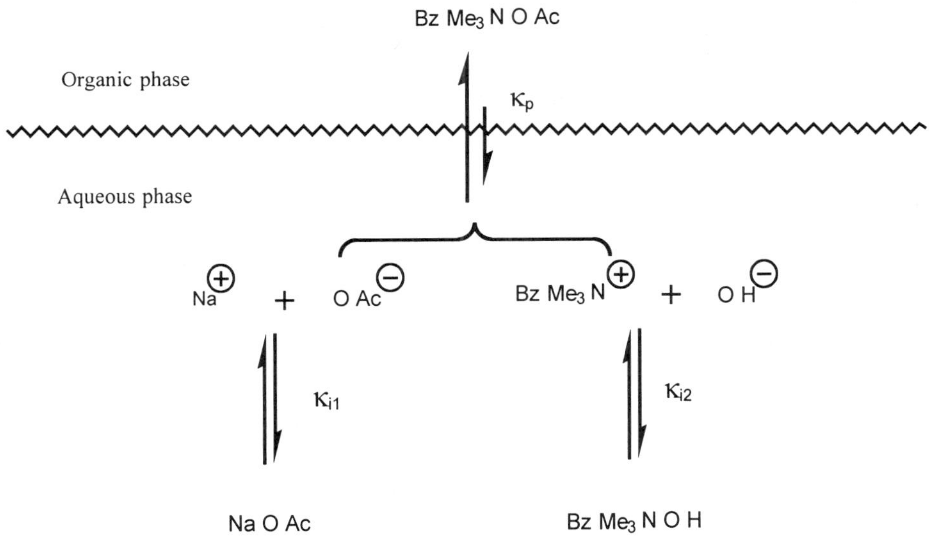

Figure 6. Effect of sodium acetate on quaternary ammonium hydroxide content in cellulose ether alkylation reactions

To evaluate the performance of the DPGE-based HMHEC polymer on paint rheology, prototypical paint formulations were prepared in the laboratory using HMHEC prepared using either sodium hydroxide (control) or with benzyltrimethylammonium hydroxide (PTC). The paint formulae using commercial latexes were supplied by latex manufacturers such as UCAR® Emulsion Systems and Rohm & Haas. The paint performance testing with these HMHEC polymers is summarized in Table 5, and standard ASTM methods were used to evaluate these paints for Stormer and ICI viscosities (D 562-81 & D 4287-88), sag resistance (D 4400-84), Leneta leveling (D 4062-88), and spatter resistance (D 4707-87). The two polymers exhibited essentially identical rheological behavior, suggesting that HMHEC polymers prepared with the same molecular weight and the same hydrophobe and hydrophobe MS are structurally identical regardless of the type of base used in their synthesis.

Table 5. Paint formulation testing of HMHEC polymers prepared with DPGE and HEC using either sodium hydroxide or phase transfer catalysis (PTC)

Latex	HMHEC polymer	Stormer viscosity	ICI viscosity	Sag resistance	Leneta leveling	Spatter resistance
Interior vinyl-acrylic flat (PVC = 60.5%)						
UCAR® latex 376	Control	93 KU	1.8 P	20	2	6
UCAR® latex 376	PTC	98 KU	1.9 P	18	2	7
Exterior all-acrylic flat (PVC = 40.0%)						
UCAR® latex 624	Control	97 KU	1.4 P	25	3	7
UCAR® latex 624	PTC	91 KU	1.4 P	25	2	8
RHOPLEX® AC-388	Control	105 KU	1.5 P	40	2	9
RHOPLEX® AC-388	PTC	104 KU	1.5 P	40	2	9

Control = prepared using NaOH; PTC = prepared using benzyltrimethylammonium hydroxide

CONCLUSIONS

Laboratory work on the heterogeneous synthesis of hydroxypropyl cellulose has shown that in the propoxylation process, the caustic and cellulose reside almost entirely in the lower aqueous phase. Our laboratory work on the hydrophobe alkylation of HEC in aqueous *t*-butyl alcohol has confirmed the same distribution of caustic and water. Because the HEC substrate is found in the aqueous phase, the rate of alkylation is partly controlled by the relative partitioning of the alkylating agent between the water and the organic phases, and the rate-limiting step in the alkylation of HEC would be the partitioning of the alkylating agent into the aqueous phase. One would therefore expect a direct relationship between water solubility and reactivity of alkylating agents with cellulose or HEC, and indeed this has been observed. A further consequence of this reaction mechanism would be the enhancement in the alkylation efficiency by the judicious use of appropriate phase transfer catalysts. In this work, the utility of quaternary ammonium hydroxides to increase the alkylation efficiency of hydrophobic reactants in the synthesis of hydrophobe-modified hydroxyethyl cellulose (HMHEC) has been demonstrated.

Of the quaternary ammonium hydroxides evaluated, benzyltrimethylammonium and tetrabutylammonium hydroxides afforded the highest increases in alkylation efficiency of dodecyl phenyl glycidyl ether (DPGE) with HEC. Maximum alkylation efficiencies were noted with a molar ratio of quaternary ammonium hydroxide to HEC of between 0.25 and 0.50. The increase in alkylation efficiency of DPGE with HEC in the presence of benzyltrimethylammonium hydroxides was observed in either aqueous *t*-butyl alcohol or aqueous acetone diluents. Depending on the specific hydrophobe, the use of benzyltrimethyl-ammonium hydroxide as a causticizing agent for HEC afforded between a two- and twelve-fold increase in the alkylation efficiency over sodium hydroxide. The largest increase in alkylation efficiency was noted with 1-bromohexadecane.

The use of mixtures of sodium hydroxide and benzyltrimethylammonium chloride or tetrabutylammonium bromide failed to enhance the DPGE alkylation of HEC by the *in situ* formation of the corresponding quaternary ammonium hydroxide phase transfer catalyst. These quaternary ammonium halides are too soluble in aqueous *t*-butyl alcohol and are preferentially extracted into the organic phase. Mixtures of benzyltrimethylammonium hydroxide and sodium acetate were also ineffective in enhancing the DPGE alkylation of HEC for the same reason, namely preferential solubility of benzyltrimethylammonium acetate in the organic phase.

The final latex paint performance properties of HMHEC polymers prepared with benzyl-trimethylammonium hydroxide appear to be the same as HMHEC polymers prepared with sodium hydroxide.

ACKNOWLEDGMENTS

The fine technical assistance of Mr. Arthur H. Marsh and Ms. Ella B. Yezrielev in conducting the polymer syntheses and paint evaluation is gratefully acknowledged. The author should also like to thank Union Carbide Corporation for permission to publish this work.

REFERENCES

1a. G. D. Shay and A. F. Rich, Urethane-functional alkali-soluble associative latex thickeners, *J. Coatings Tech.*, **58** (732): 43 (1986).
b. J. E. Hall, P. Hodgson, L. Krivanek, and P. Malizia, Influence of rheology modifiers on the performance characteristics of latex paints, *J. Coatings Tech.*, **58** (738): 65 (1986).
c. P. A. Reynolds, Association thickening of polymer latex dispersions, *Prog. Org. Coatings*, **20**: 393 (1992).
d. P. R. Howard, E. L. Leasure, S. T. Rosier, and E. J. Schaller, Formulating with combinations of HEUR associative thickeners, *J. Coatings Tech.*, **64** (804): 87 (1992).
2. G. D. Shay, Thickeners and rheology modifiers, in: *Paint and Coating Testing Manual*, J. V. Koleske, ed., 14th edition of the Gardner-Sward Handbook, ASTM, Philadelphia (1995).
3a. A. C. Sau and L. M. Landoll, Synthesis and solution properties of hydrophobically modified hydroxyethyl cellulose, in: *Polymers in Aqueous Media: Performance Through Association*, J. E. Glass, ed., ACS Advances in Chemistry Series, #223, Washington, D. C. (1989).
b. R. A. Gelman and H. G. Barth, Viscosity studies of hydrophobically modified hydroxyethyl cellulose, in: *Water-Soluble Polymers: Beauty with Performance*, J. E. Glass, ed., ACS Advances in Chemistry Series, #213, Washington, D. C. (1986).
4. H. Yokota, Alkalization mechanism of cellulose in hydroxypropylcellulose preparation process, *J. Appl. Polym. Sci.*, **32**: 3423 (1986).
5a. E. V. Dehmlow and S. S. Dehmlow, *Phase Transfer Catalysis*, 2nd. edition, Verlag Chemie, Basel (1983).
b. C. M. Starks and C. Liotta, *Phase Transfer Catalysis: Principles and Techniques*, Academic Press, New York (1978).
c. W. P. Weber and G. W. Gokel, *Phase Transfer Catalysis in Organic Synthesis*, Springer-Verlag, Berlin (1977).
d. J. Dockx, Quaternary ammonium compounds in organic synthesis, *Synthesis*, 441 (1973).
6. L. J. Mathias, Crown ethers and phase transfer catalysis in polymer chemistry, *J. Macromol. Sci., Chem.*, **A15** (5): 853 (1981).
7. J. M. J. Fréchet, Chemical modification of halogenated polymers under phase transfer conditions, *J. Macromol. Sci., Chem.*, **A15** (5): 877 (1981).
8. T. Nishikubo, T. Iizawa, K. Kobayashi, and M. Okawara, Substitution reaction of poly(chloromethyl-styrene) with some nucleophilic reagents using phase transfer catalysis, *Makromol. Chem., Rapid Comm.*, **1**: 765 (1980).
9a. R. Liotta and G. Brons, Coal. Kinetics of O-alkylation, *J. Am. Chem. Soc.*, **103**: 1735 (1981).
b. R. Liotta, K. Rose, and E. Hippo, O-Alkylation chemistry of coal and its implications for the chemical and physical structure of coal, *J. Org. Chem.*, **46**: 277 (1981).
10a. W. H. Daly and J. D. Caldwell, Influence of quaternary ammonium salts on cellulose benzylation, *J. Polym. Sci., Polym. Lett.*, **17**: 55 (1979).
b. W. H. Daly, J. D. Caldwell, K. V. Phung, and R. Tang, Influence of quaternary ammonium salts on cellulose modification, in: *Crown Ethers and Phase Transfer Catalysis in Polymer Science*, L. J. Mathias and C. E. Carraher, ed., Plenum Press, New York (1984).
11. CELLOSIZE® is Union Carbide's trademark for its hydroxyethyl cellulose (HEC) polymer.
12a. E. M. Partain, U. S. Patent 5,504,123, *Dual Functional Cellulosic Additives for Latex Compositions*, April 2, 1996.
b. E. M. Partain, U. S. Patent 5,583,214, *Dual Functional Cellulosic Additives for Latex Compositions,* December 10, 1996.
13. A. C. Cope and E. R. Trumbull, Olefins from amines: the Hofmann elimination reaction and amine oxide pyrolysis, *Organic Reactions*, **11**: 317 (1960).
14. 3-*n*-Pentadecenyl phenyl glycidyl ether is a commercial product of the Cardolite Corporation, sold under the tradename CARDOLITE® NC-513.
15a. L. Shechter, J. Wynstra, and R. E. Kurkjy, Glycidyl ether reactions with amines, *Ind. Eng. Chem.*, **48**: 94 (1956).
b. L. Shechter, J. Wynstra, and R. E. Kurkjy, Chemistry of styrene oxide, *Ind. Eng. Chem.*, **49**: 1107 (1957).
c. W. J. Belanger and J. H. Denham, The catalyzed reactions of various glycidyl ether compounds with amines, acids, phenols, and alcohols, ACS Preprints, *Division of Paint, Plastics, and Printing Ink Chemistry,* **18**(2): 10 (1958).
d. R. E. Parker and N. S. Issacs, Mechanisms of epoxide reactions, *Chem. Rev.*, **59**: 737 (1959).
16. J. Murto, Hydroxide-alkoxide ion equilibria and their influence on chemical reactions, in: *The Chemistry of Hydroxyl Groups*, S. Patai, ed. John Wiley & Sons, New York (1971).
17. E. M. Partain, G. L. Brode, and R. C. Hoy, European Patent Application 0 384 167, *Polysaccharides with Alkyl-Aryl Aromatic Hydrophobes and Latex Compositions Containing Same*, August 29, 1990.

THE SYNTHESIS AND CHARACTERIZATION OF POLYESTERS DERIVED FROM L-LACTIDE AND VARIABLY-SIZED POLY(CAPROLACTONE)

Michael R. Lostocco* and Samuel J. Huang*

*Institute of Materials Science
Polymer Program
University of Connecticut
Storrs, CT 06269-3136

INTRODUCTION

In an age of increasing societal concerns for an environmentally conscious method of disposing end-use materials, poly(lactic acid)[PLA] offers a viable means to potentially alleviate the repercussions that accompany this complex issue. With biocompatible degradation by-products and physical properties that rival those of polyethylene, polystyrene and polyvinyl chloride, PLA appears to be an attractive alternative to these high-volume commodity thermoplastics from a somewhat superficial perspective.[1] However, with its relatively high cost, PLA does not reside in the same marketability category as the aforementioned commodities. A possible resolution to this problem of economics stems from the origin of its lactic acid(LA) building blocks; that is, LA is produced from fermentation of naturally abundant carbohydrates.[2] In any event, prior to addressing the ever-present question of cost, the fact that PLA is deficient in certain areas of performance must not be overlooked. With a glass transition temperature(~55-65°C) that resides in a precarious regime if considering biodegradable, consumer product applications, less than adequate thermal and hydrolytic stability, and toughness that is highly dependent upon stereoregularity, it is obvious that much attention must be given to these limitations.

In an attempt to gain an element of control over the mechanical, physical and degradation properties of PLA, large-scale efforts are currently underway to chemically and physically modify PLA through copolymerization[3-5] and physical blending[6-12] with prudently chosen secondary constituents. Without immediate consideration for the issue of financial accessibility of PLA-based products, a logical, complementary component to PLA is poly(ε-caprolactone)[PCL]. PCL is a flexible, semi-crystalline(Tg~-60°C, Tm~60°C) aliphatic polyester with unique biodegradation characteristics and superior thermal and hydrolytic stability.[13,14]

However, while physical blending is a credible approach to altering the properties of PLA with variably-sized PCL, this method tends to neglect the specific interaction between the homopolymer and its minor components. Such neglect could prove to be

Polymer Modification
Edited by Swift *et al.*, Plenum Press, New York, 1997

detrimental to efforts designed to tailor the molecular architecture of PLA. By combining the attributes of physical blending with adequate consideration for control over interfacial interaction, reactive blending[15] offers a tool to implement such control, assuming that an appropriate combination of reaction conditions and compatibilizing agents is chosen. A prevalent technique to influence the phase morphology and mechanical performance of multiphasic polymer blends is to introduce copolymers into a system of homopolymers[16-18]. The copolymers typically resemble the individual homopolymers. In this way, the copolymers behave as 'macromolecular surfactants'[16] between phases presuming that the copolymer is able to diffuse to the interface and orient in a favorable manner under an imposed time limitation. From extensive studies upon monomer sequence length distribution and its effect upon miscibility within blends of chemically distinct polymers[16,19-35], it has been suggested that block copolymers tend to diminish the interfacial tension between phases more significantly than copolymers with more randomized structures.

Coinciding with the reported effect of block copolymers upon physical blends, this study serves as a prelude to high molecular weight PCL/PLA compatibilization by investigating the variable degree of phase separation and stability within PCL/PLA block copolyesters. By varying the size and concentration of dihydroxy-terminated PCL as the macroinitiator for l-lactide(LLA) polymerization[36], the polyesters were prepared with the intention of suppressing the ester exchange reactions that hinder the formation of block copolyesters[37,38]. As a result, by controlling the extent of these competing processes at the broad compositional range, it is proposed that certain copolyesters within the experimental range could be utilized as agents to enhance the interfacial adhesion between the respective homopolyesters.

EXPERIMENTAL

Materials: L-lactide(LLA)[Aldrich, 99+%], stannous 2-ethyl hexanoate,$Sn(Oct)_2$, catalyst[Sigma, 95%] and the purifying solvents were utilized according to the reported commercial purity. Dihydroxy-terminated PCL ranging in molecular weight from oligomeric compounds[530 and 2000(2K) Daltons] to polymeric species[43,000(43K) and 80,000(80K) Daltons] were obtained from Aldrich and Union Carbide, respectively.

Block Copolyester Formation: Each polymerization was carried out in the bulk under an inert atmosphere as summarized in Figure 3.

Prior to the inception of reaction, the reaction vessel was flame-dried and thoroughly flushed with nitrogen. For each reactive system, an increasing weight ratio of LLA to caprolactone(CL) was charged to the vessel in the presence of a catalytic amount of $Sn(Oct)_2$. With sufficient stirring provided, the vessel was immersed in an oil bath maintained at 130°C for 48 hours followed by slow cooling down to room temperature. The resulting materials were dissolved in chloroform and precipitated from an excess of hexanes or methanol(10:1 by volume) depending upon the initial molecular weight of PCL. This adjustment in purification procedure was warranted due to the tendency of those polyesters derived from oligomeric PCL to show solubility in methanol. Regardless of the specific procedure, the precipitate was dried under vacuum at room temperature for 24 hours.

Characterization: ^{1}H-NMR spectroscopy was administered using Bruker DMX-500MHz solution NMR instrumentation with $CDCl_3$ as the solvent of choice. From ^{1}H-NMR,

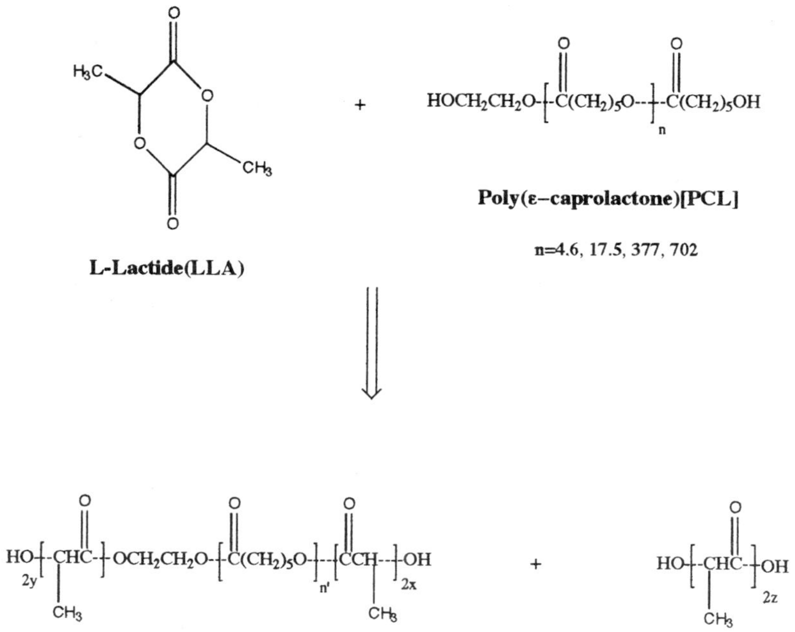

Figure 1. Reaction scheme for PCL/PLA block copolyester formation.

characteristic absorption signals facilitated a comparison between the original LLA/CL feed weight ratios and the actual ratios following purification. Differential scanning calorimetry(DSC) was performed using a TA 2920 differential scanning calorimeter. The reported thermal transitions were calculated from second heatings of the materials at 20°C per minute from -80°C to 180°C. Molecular weight analysis was conducted with the aid of gel-permeation chromatography(GPC)[Waters Model 150-C ALC] using PCL standards in tetrahydrofuran at 30°C. More subtle thermal transitions as well as the effects of LLA content upon mechanical behavior were detected with dynamic mechanical thermal analysis(DMTA)[Model DMTA MkII] at a heating of 2°C/min. from -80°C. In conjunction with DMTA results, phase morphology changes were examined with scanning electron microscopy(SEM)[Amray Model 1000A] of fracture surfaces of compression-molded films.

RESULTS AND DISCUSSION

Copolyesters from Oligomeric PCL-diols

Figures 2-5 depict the effects of CL content upon number-average molecular weight, glass transition temperature(Tg) and crystallization characteristics for those copolyesters derived from oligomeric PCL-diols.

For the LLA/530 PCL and LLA/2K PCL systems, the number-average molecular weight and Tg values are progressively reduced as the presence of CL is enhanced. These results stem from the increased availability of hydroxy-terminated species to initiate the polymerization of LLA as the amount of PCL oligomer is increased. In accord with this observation, while a simultaneous reduction of Tg with a decrease in number-average

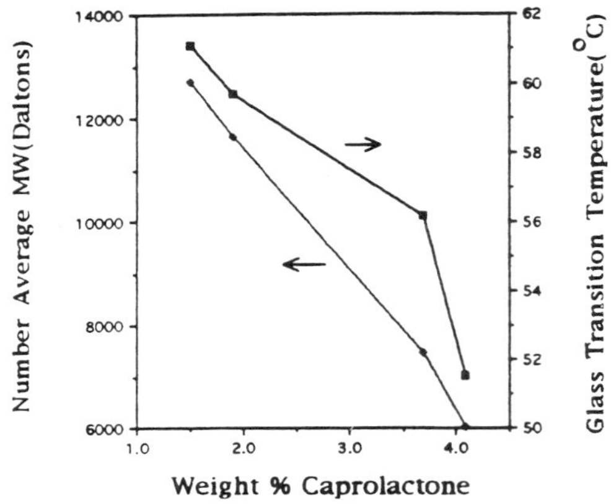

Figure 2. Effect of CL content upon number-average molecular weight and Tg of LLA/530 PCL.

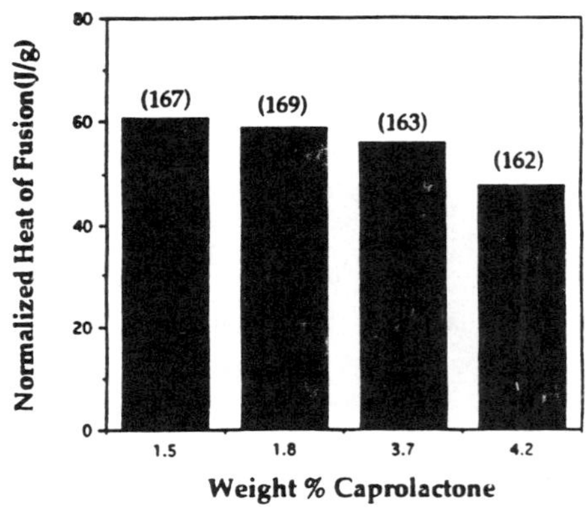

Figure 3. Heats of fusion(J/g) and (melting points) of the LLA/530 PCL systems.

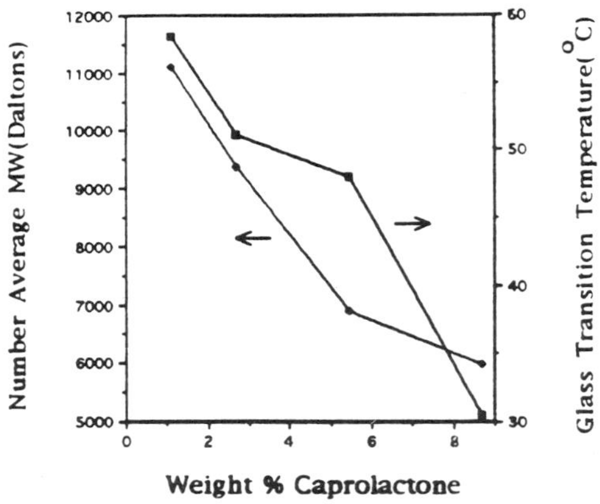

Figure 4. Effect of CL content upon number-average molecular weight and Tg of LLA/2K PCL.

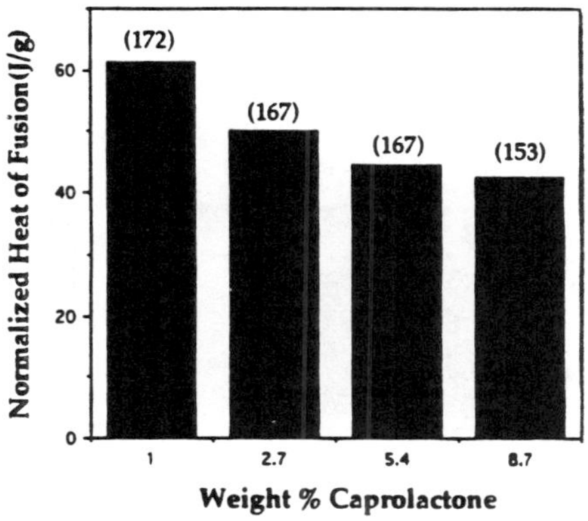

Figure 5. Heats of fusion(J/g) and (melting points) of the LLA/2K PCL systems.

molecular weight is anticipated, this effect may be accentuated by the sequence length and/or distribution of flexible, CL units. This point may be illustrated with the aid of Figures 2 and 4. If similar CL compositions are compared, the Tg values of the LLA/2K PCL copolyesters are reduced to a greater extent than those observed for the LLA/530 PCL systems. This effect becomes more pronounced for the LLA/2K PCL copolyesters beyond ~4%CL.

With regards to the crystallization behavior of these copolyesters(Figures 3 and 5), a single melting peak indicative of a PLLA crystalline phase is seen. The fact that the PCL phase does not exhibit the ability to crystallize following the polymerization of LLA may be attributed to the occurrence of ester exchange reactions that limit the average sequence

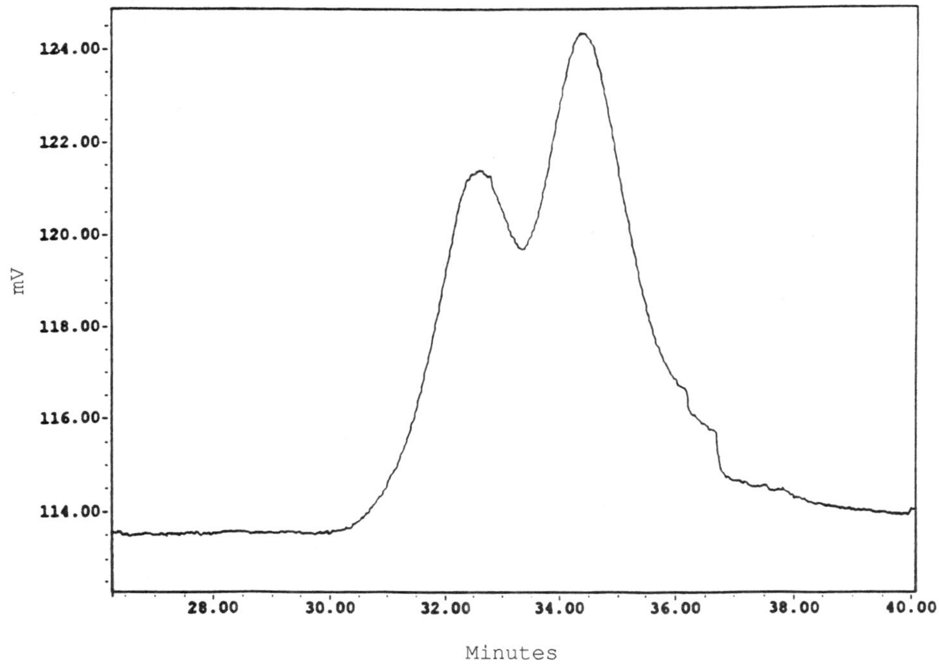

Figure 6. Representative gel-permeation chromatograph of copolyesters derived from oligomeric PCL.

Table 1. Synthetic, molecular weight and thermal properties of LLA/43K and LLA/80K PCL copolyesters.

Polyester	*LLA/CL*[♣]	*LLA/CL*[♦]	*%Yield*[a]	*Mn*[b]	*PDI*[b]	*Tg(°C)*[c]	*Tm(°C)*[c]
LLA/43K	0.26	0.20(**83.3**)	84	25200	1.5	-----	57, 168
LLA/43K	0.42	0.34(**74.6**)	83	27500	1.4	-----	57, 171
LLA/43K	0.37	0.77(**56.5**)	77	25600	1.4	-----	55, 171
LLA/43K	0.69	1.85(**35.1**)	75	27000	1.4	-----	56, 172
LLA/80K	0.14	0.03(**97.1**)	78	33900	1.3	-----	58, 167
LLA/80K	0.23	0.13(**88.5**)	83	25200	1.5	-----	56, 168
LLA/80K	0.46	0.34(**74.6**)	81	31900	1.3	-----	56, 168
LLA/80K	0.87	1.11(**47.4**)	80	32100	1.3	-----	54, 170

[♣]Initial feed weight ratio of LLA/CL.
[♦]Obtained from ^{1}H-NMR spectroscopy utilizing $CDCl_3$/TMS(**%CL**).
[a]Calculated following $CHCl_3$ dissolution and precipitation from methanol.
[b]Acquired from GPC using PCL standards in tetrahydrofuran @30°C.
[c]Reported values from DSC measurements[2nd heatings].

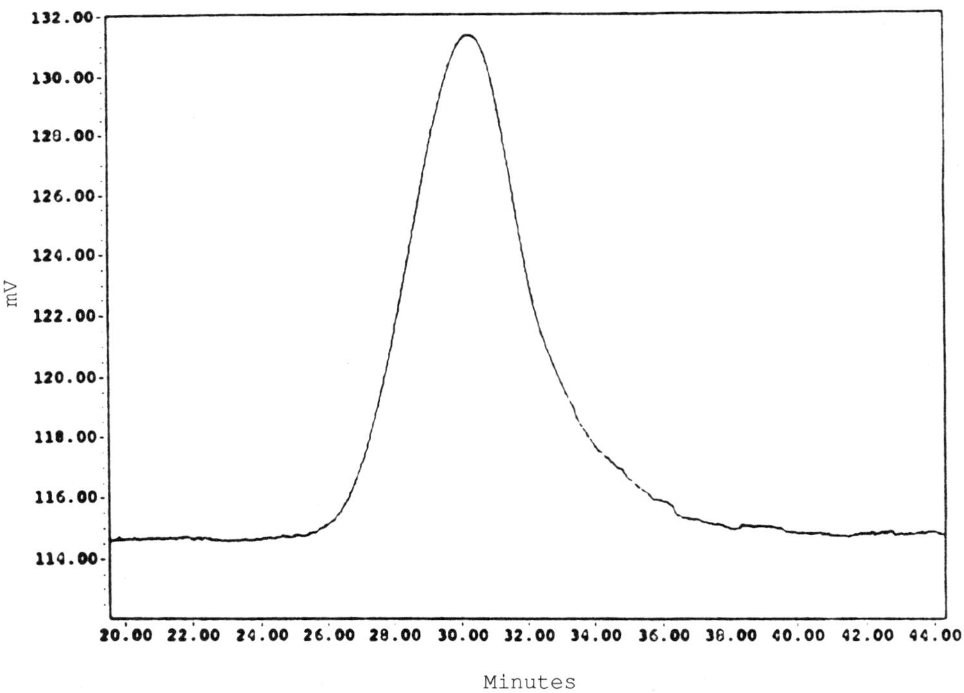

Figure 7. Representative gel-permeation chromatograph of copolyesters derived from higher MW PCL.

length of CL beyond the critical length required for chain folding.[39] In any event, as CL is incorporated into the LLA/530 PCL and LLA/2K PCL copolyesters, the melting points and normalized heats of fusion decrease as a consequence of more significant alterations in the resultant crystal structures. However, since concentration may play a crucial role in crystallization kinetic and thermodynamic parameters, judgments regarding specific architectural effects must be reserved until actual sequence lengths are determined.

Finally, chromatographic elution characteristics(Figure 8) further complicate any definitive explanation of CL content and its effect upon amorphous and crystalline phase modifications. With the appearance of bimodal elution, the oligomeric component of the copolyesters may act as a plasticizing agent to the PLLA glassy state not withstanding its specific chemical composition (i.e.-exchanged copolyester or PLLA oligoester).

Copolyesters Derived from Higher Molecular Weight PCL-diols

Table 1 outlines the synthetic characteristics, molecular weight trends and thermal properties of those copolyesters fabricated from 43K and 80K PCL-diols and LLA. Similar to the LLA/530 PCL and LLA/2K PCL systems, in accord with the reported % yield values, the discrepancy between initial charge weight ratios of LLA/CL and actual ratios calculated from ^{1}H-NMR spectroscopy following reaction may be ascribed to the removal of lower molecular weight species during purification. Contrary to the LLA/530 PCL and LLA/2K PCL materials, with unimodal elution observed(Figure 7), the molecular weights of the LLA/43K and LLA/80K PCL copolyesters consistently approach ~25,000-34,000 Daltons due to the degradative presence of alcoholysis and hydrolysis reactions. Any interpretation of the less disperse molecular weight distributions for these systems may originate from the reduced concentration of hydroxy-terminated PCL molecules that are imperative in the process of ring-opening polymerization of LLA. Also, with only two clear melt endotherms seen at 54-58°C for PCL and 168-172°C for PLLA, these polyesters do not reveal the same semi-crystalline behavior. Figures 8 and 9 compare the ability of the PCL and PLLA phases to crystallize as a function of LLA/CL weight ratios. In each instance, the fractional crystallinity of the PLLA phase increases with an increase in the LLA content with a concomitant reduction in the PCL phase crystallinity. Nevertheless, in order to thoroughly understand the multiphasic character of these copolyesters, it must be realized that the sensitivity limitations of DSC may prevent a comprehensive analysis of each thermal transition of the respective components.

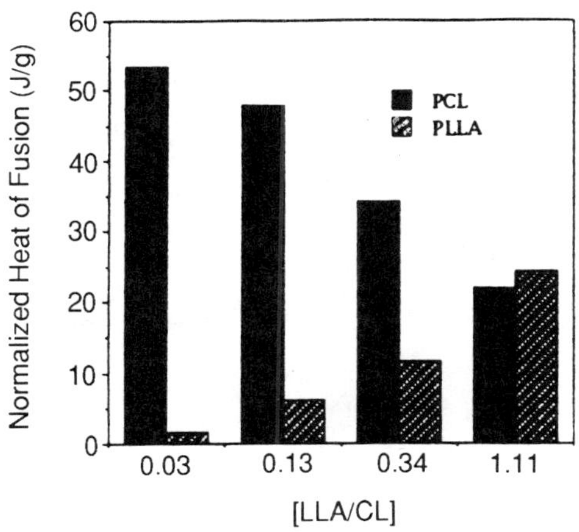

Figure 8. Relative normalized heats of fusion(J/g) within the LLA/43K PCL copolyesters.

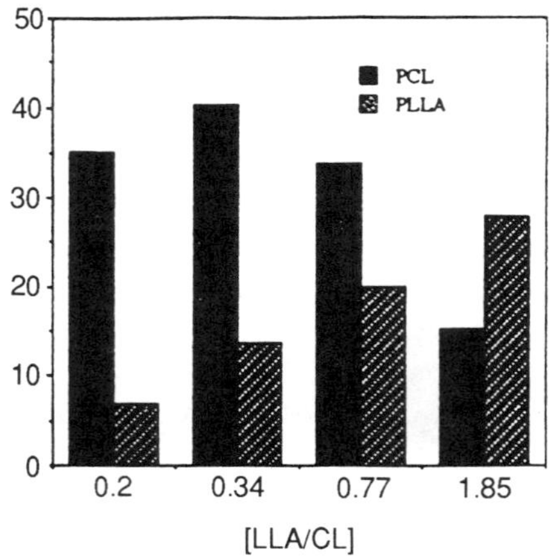

Figure 9. Relative normalized heats of fusion(J/g) within the LLA/80K PCL copolyesters.

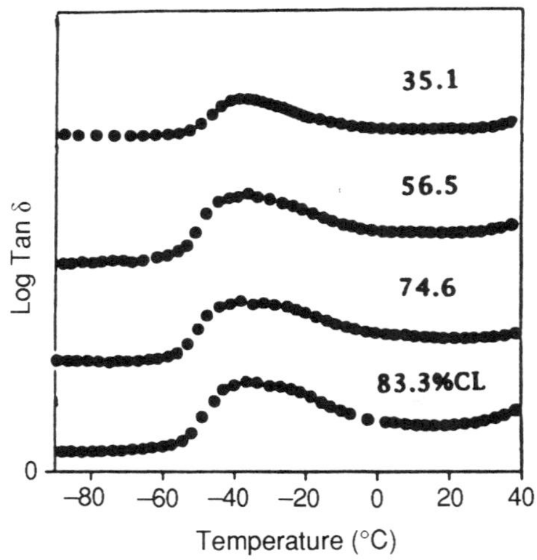

Figure 10. Log tan δ vs. temperature(°C) for LLA/43K PCL systems.

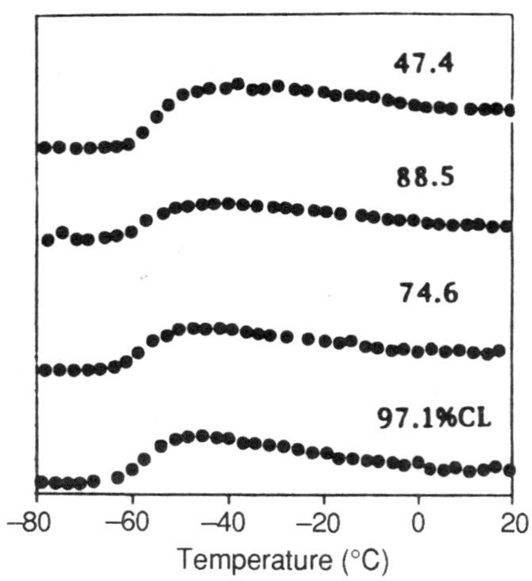

Figure 11. Log tan δ vs. temperature(°C) for LLA/80K PCL systems.

This theory has been verified with respect to the PCL amorphous phase. Figures 10-13 focus upon the dynamic mechanical thermal properties of the copolyesters under sub-ambient conditions.

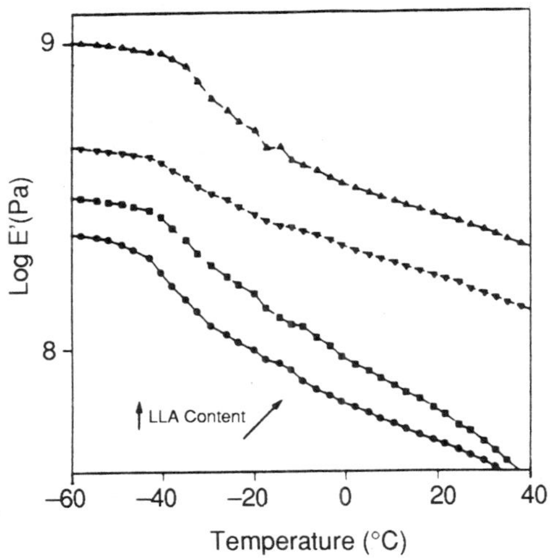

Figure 12. Log E'(Pa) vs. temperature(°C) for the LLA/43K PCL systems.

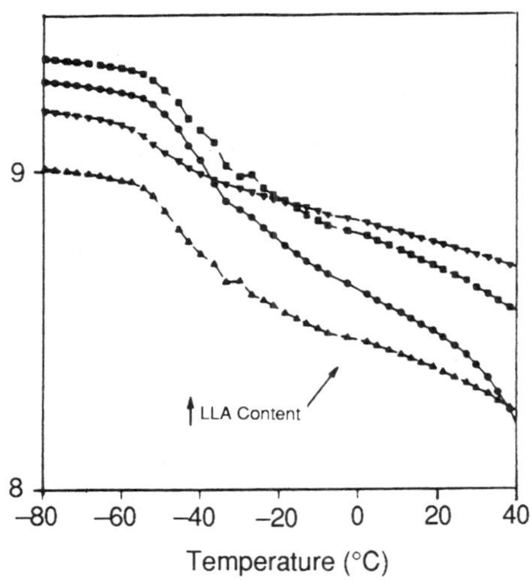

Figure 13. Log E'(Pa) vs. temperature(°C) for the LLA/80K PCL systems.

Figures 10 and 11 show successive log tan δ curves with increasing LLA content. The peak maxima do not diverge from ~-40°C for the LLA/43K PCL copolyesters and ~-50°C for the LLA/80K PCL copolyesters, implying that a higher degree of amorphous compatibility exists within the LLA/43K PCL systems. In addition, when considering the magnitude of the tan δ peaks, as the PLLA crystalline phase is augmented, the corresponding damping functions tend to flatten.[40] These effects are further substantiated through log E' versus temperature diagrams(Figures 12 and 13). The inflection points

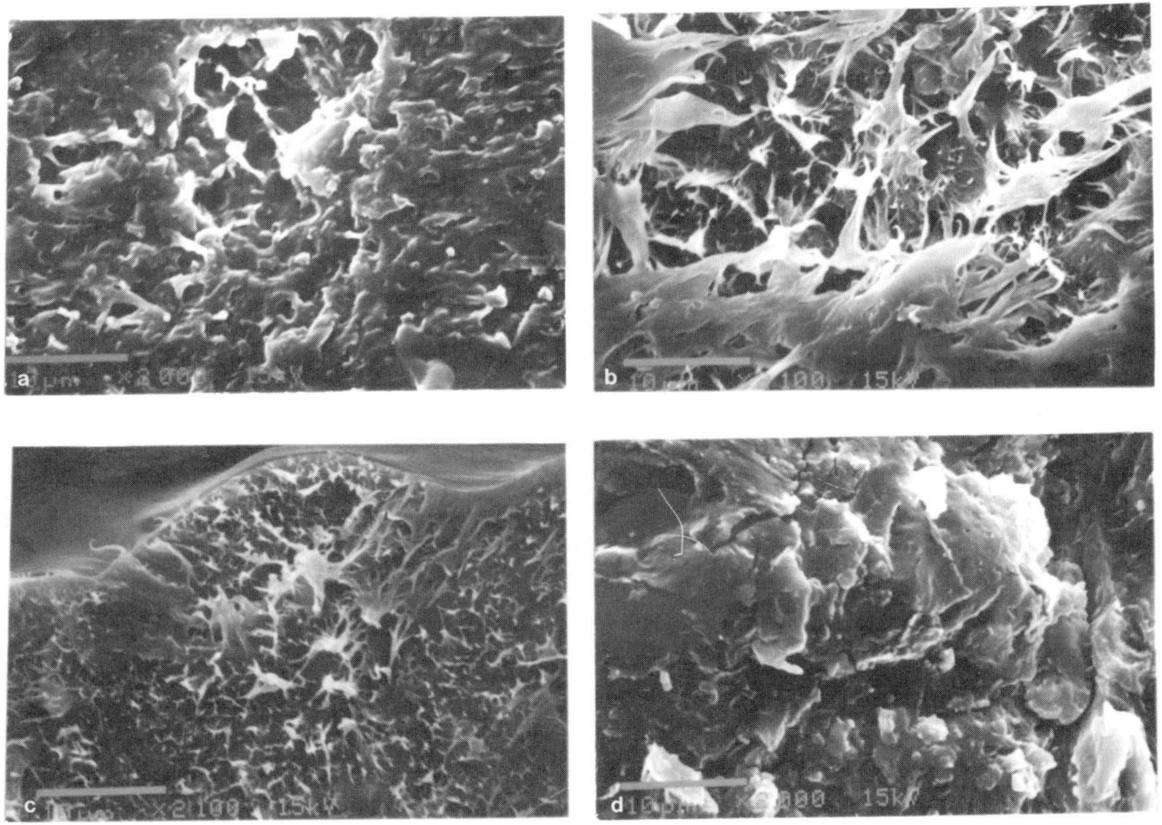

Figure 14. Scanning electron micrographs of LLA/43K PCL copolyesters with defined magnification: a.) 83.3%CL(2000X), b.) 74.6%CL(2100X), c.) 56.5%CL(2100X), d.) 35.1%CL(2000X).

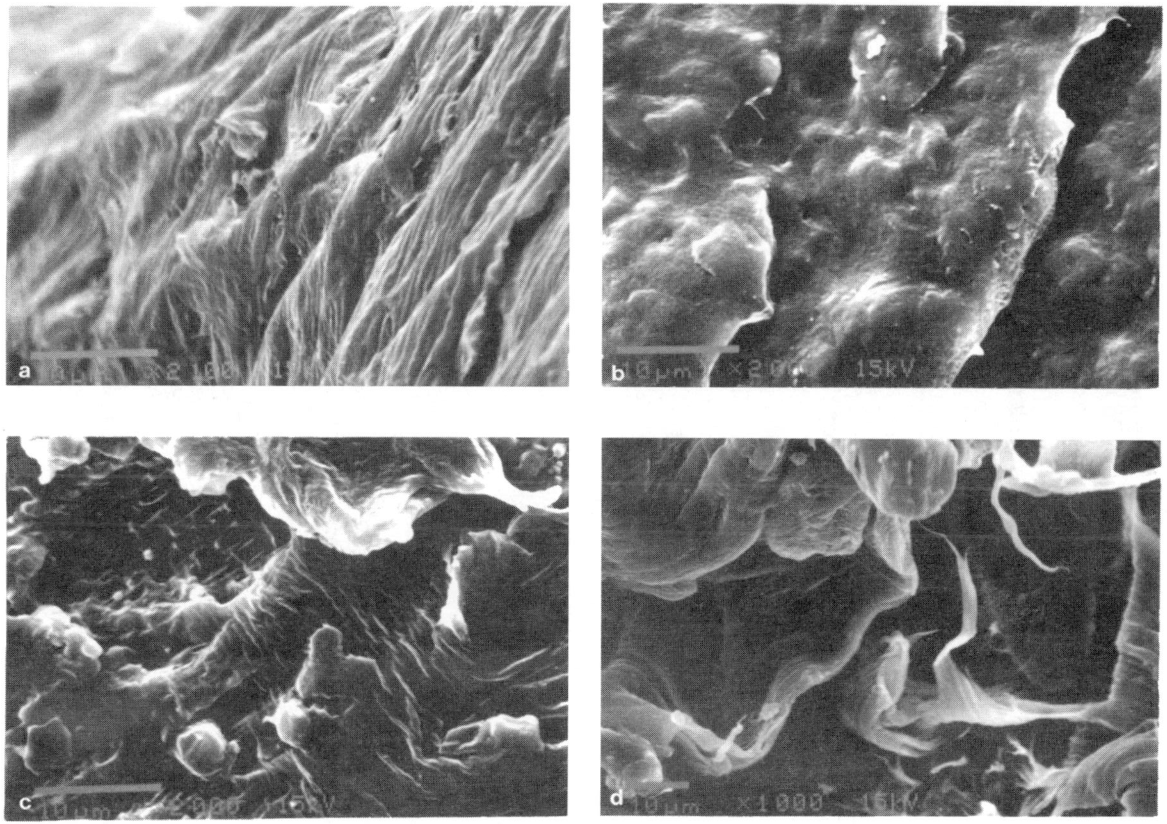

Figure 15. Scanning electron micrographs of LLA/80K PCL copolyesters with defined magnification: a.) 97.1%CL(2100X), b.) 88.5%CL(2000X), c.) 74.6%CL(2000X), d.) 47.4%CL(1000X).

representative of the PCL amorphous phase are shifted slightly towards the expected Tg value of the PLLA phase. The PLLA crystallinity appears to reinforce the copolyesters as the moduli within the rubbery plateau region following inflection are amplified. In any event, uncertainty remains regarding the exact location of the PLLA glass transition.

From a morphological standpoint, Figures 14 and 15 support the notion that the PLLA phase most probably promotes stiffening of the copolyesters beyond the Tg of PCL. Throughout each concentration range, the materials undergo structural transformations as denoted through magnification of the fracture surfaces. The striations that may result from thermal processing at 97.1% CL(Figure 15a) do not appear within the lower concentrations of CL as a more 'threadlike' texture becomes evident with reasonably good adhesion between phases. Further still, the surface appearance changes to a less fibrillar morphology(Figure 14d) as the fraction of LLA predominates.

CONCLUSIONS

As a prelude to an investigation concerning PCL/PLA blend compatibilization, a series of copolyesters derived from l-lactide and variably-sized PCL have been synthesized. Depending upon the initial molecular weight of the PCL macroinitiator, control may be attained with respect to chain size as well as amorphous and crystalline phase separation. This multiphasic character may be directly correlated to the average sequence length/distribution of the copolyester residues. Moreover, the l-lactide within the copolyesters seems to reinforce the higher molecular weight materials as manifested through dynamic mechanical thermal properties and morphological alterations. Such changes could prove to be beneficial as these copolyesters are incorporated into blends of PCL and PLA with the hope of adjusting the interfacial tension between the homopolyesters and thus, provide a wide range of mechanical properties, thermal and hydrolytic stability, and enzymatic degradability.

ACKNOWLEDGEMENT

Financial support from the National Corn Growers Association and the State of Connecticut Critical Technology Funds are gratefully acknowledged.

REFERENCES

1.) R.G. Sinclair, *ACS Abstr. Polym. Mat. Sci. Eng.*(1995), **72**, 133.
2.) L. Stryer, Biochemistry, 3rd Edition(1988), W.H. Freeman and Co., N.Y., pp. 330.
3.) A. Schindler, C.G. Pitt, R. Jeffcoat, R.A. Zweidinger, *J. Biomed. Mat. Res.*(1979), **13**, 497.
4.) X.D. Feng, C.X. Song, W.Y. Chen, *J. Polym. Sci.: Polym. Lett. Edn.*(1983), **21**, 593.
5.) M. Bero, G. Adamus, J. Kasperczyk, H. Janeczek, *Polymer Bulletin*(1993), **31**, 9.
6.) Y. Cha, C.G. Pitt, *Biomaterials*(1990), **11**, 108.
7.) H.M. Dollinger, S.P. Sawan, *Polymer Preprints*(1990), **32**, 429.
8.) A. Gajria, V. Dave, R.A. Gross, S.P. McCarthy, *Antec Proceedings*, Boston, 1995.
9.) S.P. McCarthy, C.L. Yue, V. Dave, R.A. Gross, *Polymer Preprints*(1995), **36**, 418.
10.) S. Iannace, L. Ambrosio, S.J. Huang, L. Nicolais, *J. Appl. Polym. Sci.*(1994), **54**, 1525.
11.) S. Iannace, L. Ambrosio, S.J. Huang, L. Nicolais, *J. Pure Appl. Chem.*(1995), **A32**(4), 881.

12.) L. Zhang, C. Xiong, X. Deng, *J. Appl. Polym. Sci.*(1995), **56**, 103.
13.) C.V. Benedict, W.J. Cook, P. Jarrett, J.A. Cameron, S.J. Huang, J.P. Bell, *J. Appl. Polym. Sci.* (1983), **28**, 327.
14.) C.V. Benedict, J.A. Cameron, S.J. Huang, *J. Appl. Polym. Sci.*(1983), **28**, 335.
15.) A.Y. Coran, *Rubber Chem. Technology.*(1990), **63**, 599.
16.) K.R. Shull, A.J. Kellock, V.R. Deline, S.A. MacDonald, *J. Chem. Phys.*(1992), **97**, 2095.
17.) Z. Liang, H.L. Williams, *J. Appl. Polym. Sci.*(1991), **43**, 379.
18.) S. Thomas, R. Asaletha, M.G. Kumaran, *Rubber Chem. Technology*(1995), **68**, 671.
19.) C.F. Hammer, *Macromolecules*(1971), **4**, 69.
20.) G.A. Zakrzewski, *Polymer*(1973), **14**, 347.
21.) D.J. Stein, R.H. Jung, K.H. Illers, H. Hendus, *Angew. Makromol. Chem.*(1974), **36**, 89.22.) L.P. McMaster, *Adv. Chem. Ser.*(1975), **143**, 43.
23.) P.R. Alexandrovich, F.E. Karasz, W.J. MacKnight, *Polymer*(1977), **18**, 1022.
24.) J.R. Fried, F.E. Karasz, W.J. MacKnight, *Macromolecules*(1978), **11**, 150.
25.) J.S. Chion, D.R. Paul, J.W. Barlow, *Polymer*(1982), **23**, 1543.
26.) S.H. Goh, D.R. Paul, J.W. Barlow, *Polym. Eng. Sci.*(1982), **22**, 34.
27.) R. Vukovic, F.E. Karasz, W.J. MacKnight, *Polymer*(1983), **24**, 529.
28.) R. Vukovic, V. Kuresevic, N. Segudovic, F.E. Karasz, W.J. MacKnight, *J. Appl. Polym. Sci.*(1983), **28**, 1379.
29.) S.C. Chiu, T.G. Smith, *J. Appl. Polym. Sci.*(1984), **29**, 1797.
30.) Y. Maeda, F.E. Karasz, W.J. MacKnight, R. Vukovic, *J. Polym. Sci., Part B*(1986), **24**, 2345.
31.) K.E. Min, D.R. Paul, *J. Appl. Polym. Sci.*(1989), **37**, 1153.
32.) H.I. Kim, E.M. Pearce, T.K. Kwei, *Macromolecules*(1989), **22**, 3498.
33.) J.H. Kim, J.W. Barlow, D.R. Paul, *J. Polym. Sci., Part B*(1989), **27**, 2211.
34.) J. Kressler, H.-W. Hammer, U. Morgensten, B. Litauszki, W. Berger, *Makromol. Chem.*(1990), **191**, 243.
35.) J.M.G. Covois, I.J. McEwen, L. Nadvornik, *Macromolecules*(1990), **23**, 5106.
36.) M. Spinu, C. Jackson, M.Y. Keating, K.H. Gardner, *J. Pure Appl. Chem.*(1996), **A33**(10), 1497.
37.) D.W. Grijpma, A.J. Pennings, *Polymer Bulletin*(1991), **25**, 335.
38.) J. Kasperczyk, M. Bero, *Makromol. Chem.*(1993), **194**, 913.
39.) P. Jarrett, *Ph.D. Dissertation in Polymer Science(1983): The Morphology and Mechanism of the Biodegradation of Polycaprolactone*, Institute of Materials Science, University of Connecticut, Storrs, CT.
40.) R.W. Gray, N.G. McCrum, *J. Polym. Sci.*(1969), **A2**(7), 1329.

REACTIONS WITH VINYL POLYMERS

THE FUNCTIONALIZATION OF POLYOLEFINS BY USING REACTIVE INTERMEDIATES

T. C. Chung

Department of Materials Science and Engineering,
The Pennsylvania State University
University Park, PA 16802

INTRODUCTION

Despite the commercial successes of polyolefins [1], the lack of compatibility with other materials has limitted some of their end uses. Accordingly, the chemical modification of polyolefins has been an area of increasing interest as a route to higher value products, and various methods of functionalization [2-4] have been employed to alter their chemical and physical properties. An established technique for improving the interfacial interaction between polymers and other materials is the use of block and graft copolymers as compatibilizers [5,6]. Unfortunately, the chemistry to prepare polyolefin graft and block copolymers are also very limited. Numerous methods have been employed in forming graft copolymers with polyolefins. Ionizing radiation (x-ray, γ-rays, and e-beams), ozone, uv with accelerators, and free radical initiators in the presence of monomers [7,8] have all been used to form graft copolymers. Typically, these high energy reactions lead to side reactions such as crosslinking and chain cleavage resulting in diminished mechanical properties. In most cases, the structure and composition of copolymers are difficult to controlled with the considerable amounts of ungrafted homopolymers.

BORANE COMONOMER / RADICAL APPROACH

Equation 1 illustrates a synthetic scheme which we have used in the preparation of polyolefin graft copolymers. Borane group containing polyolefin copolymers (I) were obtained by direct copolymerization [9,10] of α-olefin/borane containing α-olefin and a

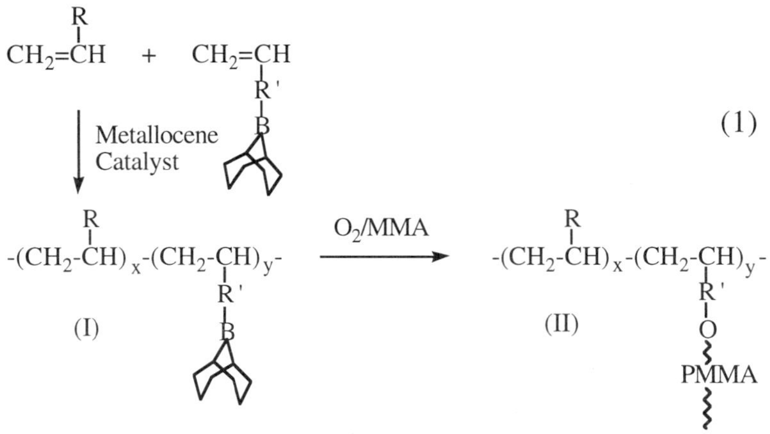

(1)

metallocene catalyst, such as $Et(Ind)_2ZrCl_2$/MAO with strained ligand geometry. The borane groups in polyolefin (I) acts as the free radical sources for the graft-from reactions. Under certain oxidative conditions, the linear alkyl C-B bond is selectively oxidized [11] to produce peroxyborane (C-O-O-B) which behaves very differently from regular benzoyl peroxides and consequently decomposes at ambient temperature. The decomposition reaction basically follows the homolytical cleavage of peroxide to generate an alkoxy radical (C-O∗) and a borinate radical (B-O∗). The alkoxyl radical is active in initiating polymerization and the borinate "dormant" radical form a reversible bond with radical at the growing polymer chain end to prolong the life time of the propagating radical. In fact, the stabilization mechanism of the borinate radical (B-O∗) is just the mirror image of nitroxide radicals, such as 2,2,6,6-tetramethyl-1-piperidinyloxy (TEMPO) radical as illustrated below:

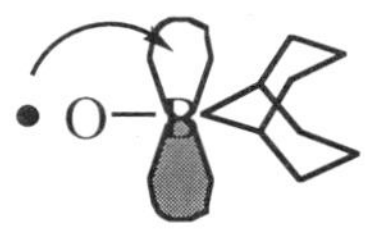

Borinate

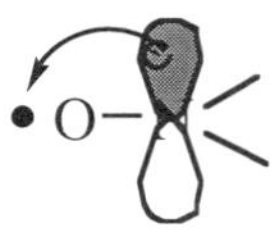

Nitroxide

The free radical polymerized functional polymers with controllable molecular weight are therefore chemically bonded to the side chains of polyolefin (II). One example of PP-g-PMMA [12,13] is shown in Table 1.

Table 1. A Summary of PP-g-PMMA Copolymers.

Run	Mole % Borane in PP	O_2 (ml/hr.)	Monomer/ Solvent	Reaction Time (hrs.)	Mole % MMA in Polymer
A-1	0.5	1.5/12	MMA (neat)	48	66
A-2	0.5	3.0/1	MMA (neat)	2	6
A-3	0.5	1.4/3	MMA/THF	12	52
A-4	0.5	6 (at once)	MMA (neat)	48	1.5
A-5	0.5	diffusion	MMA/THF	48	12

The comparison among runs (A-1 to A-4) shows the sensitivity of oxygen addition to the graft efficiency. The best results in this heterogeneous reaction system are realized when the O_2 is introduced slowly so that O « B at any time.

P-MS COMONOMER / ANIONIC APPROACH

Equation 2 illustrates another route to prepare graft copolymers, involving the preparation of polyolefin containing para-methylstyrene groups (I').

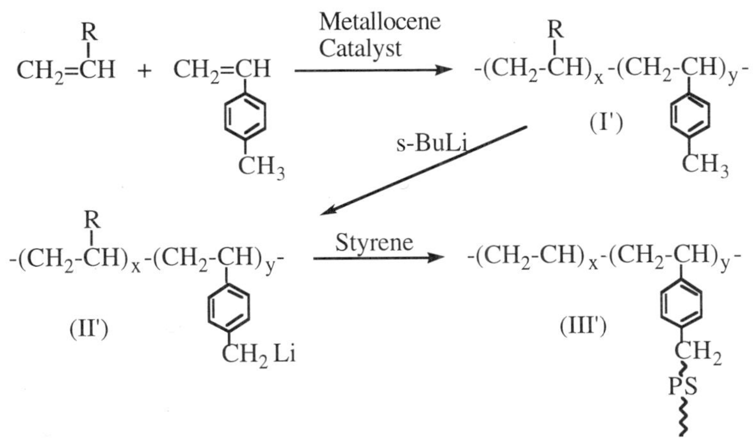

(2)

The metallocene catalyst with cationic nature and spatially opened active site provides favorable condition for the incorporation of p-alkylstyrene (p-ms) to polyolefins. The p-ms groups can be easily metallated to produce "stable" polymeric anions for graft-from polymerization. With the coexist of anion-polymerizable monomers, we have prepared many graft copolymers, such as PE-g-PS, PE-g-PMMA, PE-g-PAN, PP-g-PS, PP-g-PB, PP-g-PI and PP-g-PMMA.

Copolymerization of Ethylene and P-methylstyrene

In the ethylene copolymerization reaction, it is logical to predict that p-methylstyrene may show significantly better incorporation than styrene by using metallocene catalysts, such as $[C_5Me_4(SiMe_2N^tBu)]TiCl_2$ or $Et(Ind)_2ZrCl_2$, with constrained ligand geometry. It is well-accepted that the cationic coordination mechanism [14] is responsible for the metallocene polymerization reaction. The electron-donating from p-methyl group provides favorable electronic effect to increase the monomer reactivity. In fact, some experimental results [15] have shown the significantly higher reactivity of p-methylstyrene relative to styrene in homo-polymerization reactions using syndiotactic specific metallocene catalyst. In addition, it has also been reported that p-methylstyrene and isobutylene have similar reactivity ratios under cationic polymerization conditions. An entire compositional range of poly(isobutylene-co-p-methylstyrene) copolymers [16] have been prepared.

Table 2 summarizes the experimental results [17], which includes two sets of comparative experiments (runs 2-5 and 6-9) and a control homo-polymerization (run 1).

Table 2. A summary of copolymerization reactions between ethylene (M_1) and styrene derivatives (M_2)

Reaction Condition[1]				Copolymer Product			
Temp. (°C)	Cat.[2] (umol)	Solvent	Comonomers[3] M_1(psi) / M_2(mmol)	Yield (g)	M_2 Conc. (mole%)	M_2 (%) conversion	Tm (°C)
30	(I) 10	toluene	45 / none	4.27	0	0	133.7
30	(I) 10	toluene	45 / p-MS (46.6)	13.0	11.0	84.3	76.0
30	(I) 10	toluene	45 / o-MS (46.6)	12.9	4.52	38.7	98.3
30	(I) 10	toluene	45 / m-MS (46.6)	5.43	2.36	9.53	119.1
30	(I) 10	toluene	45 / styrene (46.6)	13.6	5.35	48.6	98.7
53	(II) 17	hexane	43.5 / p-MS (33.9)	24.7	3.30	81.5	114.3
53	(II) 17	hexane	43.5 / o-MS (33.9)	23.0	2.54	63.5	118.6
53	(II) 17	hexane	43.5 / m-MS (33.9)	19.0	2.35	43.2	118.1
53	(II) 17	hexane	43.5 / styrene (33.9)	17.5	1.87	31.7	127.4

[1] reaction time = 1 hour.
[2] $[C_5Me_4(SiMe_2N^tBu)]TiCl_2$ (I) and $Et(Ind)_2ZrCl_2$ (II) Catalysts
[3] M_1: ethylene; p-MS: p-methylstyrene; o-MS: o-methylstyrene; m-MS: m-methylstyrene.

Each set of experiments was carried out under the same reaction condition except using different comonomers, i.e. p-methylstyrene, o-methylstyrene, m-methylstyrene and styrene, respectively. The compositions of copolymers were determined by ^{1}H NMR spectra, and the thermal properties (melting point and crystallinity) were obtained by DSC measurements. Overall, all comonomers show no retardation to the catalyst activity. In fact, the significantly higher catalyst activities were observed in all copolymerization reactions (runs 2-5), comparing with that of ethylene homopolymerization (run 1). Within each set (runs 2-5 and 6-9) of comparative experiments, p-methylstyrene consistently shows better incorporation than the rest of comonomers, i.e. o-methylstyrene, m-methylstyrene and styrene. Both catalysts with constrained mono- and di-cyclopentadienyl ligands are very effective to incorporate p-methylstyrene into polyethylene backbone. In runs 2 and 6, more than 80 % of p-methylstyrene were converted to copolymer within one hour under constant (~ 45 psi) ethylene pressure. On the other hand, only less than half of styrenes (runs 5 and 9) were incorporated into ethylene copolymers under the same reaction conditions. The significantly lower styrene incorporation may be due to the lack of electronic donating from p-methyl substitution which is very favorable in the "cationic" polymerization mechanism. On the other hand, both isomers (o-methylstyrene and m-methylstyrene) were not receiving the full benefits of the electronic and steric effects existed in p-methylstyrene. Some steric hindrance of the methyl group to the propagating site may affect the incorporation of o-methylstyrene. Both less favorable electronic and steric effects may contribute significantly to the low incorporation of m-methylstyrene.

Microstructure of Poly(ethylene-co-p-methylstyrene)

The poly(ethylene-co-p-methylstyrene) with 10.9 mole % of p-methylstyrene shows a low melting point (76 °C) and very small crystallinity (5.4%), which implies the random distribution of p-methylstyrene along the polyethylene backbone [17]. The detailed sequence distribution can be quantitatively determined by ^{13}C NMR measurements.

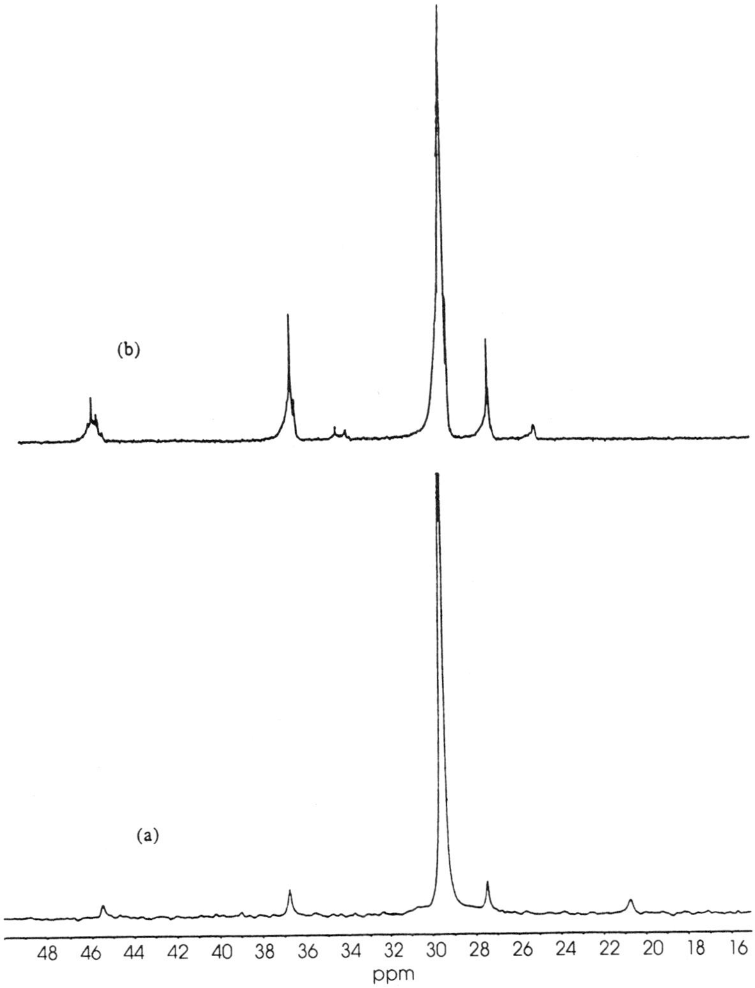

Figure 1. The ^{13}C NMR spectra of (a) poly(ethylene-co-styrene) and (b) poly(ethylene-co-p-methylstyrene), containing (a) 9.5 mole % of styrene and 10.9 mole % of p-methylstyrene.

Figure 1 compares the expanded ^{13}C NMR spectra of poly(ethylene-co-p-methylstyrene) and poly(ethylene-co-styrene), containing about 10 mole% of comonomers (i.e. p-methylstyrene and styrene).It's logical to expect that the methyl group substitution at the para-position will have very little effect on the chemical shifts of methylene and methine carbons in the polymer backbone. In other words, the comparison of aliphatic chemical shifts is a direct comparison of their sequence distributions between poly(ethylene-co-p-methylstyrene) and poly(ethylene-co-styrene). In general, fewer chemical shifts shown in the poly(ethylene-co-p-methylstyrene) sample imply more homogeneous copolymer microstructure. For detailed analysis, the experimental results were compared with the literature reference [18] and the theoretical chemical shifts which were calculated based on the improved Grant and Paul empirical method [19]. Table 3 shows the summary of the calculated and observed chemical shifts for methylene and methine carbons in the polymer backbone.

Table 3. The comparison of calculated and observed chemical shifts for methylene and methine carbons in poly(ethylene-co-p-methylstyrene) and poly(ethylene-co-styrene).

Carbon type[1]	Sequence[2]	Calculated (ppm)	Observed P(E-co-p-MS) (ppm)	Observed P(E-co-S) (ppm)
$S_{\beta\beta}$	SES	25.91		25.22
$S_{\beta\delta}$	SEE	28.01	27.74	27.36
$S_{\delta\delta}$	EEE	30.13	29.80	29.26
$S_{\gamma\delta}$	EEE	30.56		29.42
$S_{\alpha\alpha}$	SSE	34.75		34.13 / 34.60
$S_{\alpha\delta}$	SES	37.10	37.04	36.48
$S_{\alpha\gamma}$	SEE	37.10		36.65
$T_{\beta\beta}$	SSS	41.60		
$T_{\beta\delta}$	SSE	45.53		45.64
$T_{\delta\delta}$	ESE	47.60	45.77	46.08

[1]S: methylene carbon; T: methine carbon; α,β, γ, δ: distance from methylene or methine carbon to the adjacent aromatic substituted methine carbon. (δ includes the distance beyond four carbons)

[2]E: ethylene; S: styrene or p-methylstyrene

In general, the calculated and experimental results are in good agreement. The small deviation in the exact chemical shift is most likely due to the treatment of the phenyl side group which was not substantiated from the alkane group in the Grant and Paul empirical method. Every chemical shift in Figure 1 (a) can be clearly assigned. In addition to the two chemical shifts (21.01 and 29.80 ppm), corresponding to the methylene carbons from ethylene and methyl carbons from p-methylstyrene, respectively, there are three well-resolved peaks (27.74, 37.04 and 45.78 ppm) corresponding to methylene and methine carbons from p-methylstyrene units which are separated by multiple ethylene units along the polymer chain. On the other hand, the spectrum in Figure 1 (b) shows much more complicated methylene and methine carbon species in poly(ethylene-co-styrene) sample with several relatively broad bands. Many consecutive and adjacent styrene units (SSE and SES) clearly exist in the polymer chain, as shown and explained in Table 3. There is no detectable SSS sequence in the copolymer.

Anionic Graft-from Reactions

Our major research interest of incorporating p-methylstyrene into polyolefins is due to its versatility to access a broad range of functional groups. The benzylic protons are ready for many chemical reactions, such as halogenation, metallation and oxidation, which can introduce functional groups at the benzylic position under mild reaction conditions. In addition, the anionic graft-from reaction offers a relatively simple process in the preparation of graft copolymers. The lithiated polymer was suspended in an inert organic diluent before addition of monomers, such as styrene, MMA, vinyl acetate, acrylonitrile and p-methylstyrene, the living anionic polymerization takes place as well-known solution anionic polymerization. It is important to note that the anionic polymerization of various monomers, such as methyl methacrylate, can take place at room temperature without causing any detectable side reactions, which may be associated with the stable benzylic anion in solid form. After achieving the desirable composition of the graft copolymer, the graft-from reaction was terminated by an alcohol, such as methanol. Good solvents for backbone and side chain polymers were used during the fractionization, using a Soxhlet apparatus under N_2 for 24 hours. The soluble fractions were isolated by vacuum-removal of solvent. Usually, the total soluble fractions were less than 5 % of the product. The major insoluble fraction was PE graft copolymer which was completely soluble in xylene or trichlorobenzene at elevated temperatures.

Figure 2 shows the ^{1}H NMR spectra of three PE-g-PS copolymers. Comparing with the ^{1}H NMR spectrum of the starting PE-co-(p-MS), three additional chemical shifts show around 1.55, 2.0 and 6.4-7.3 ppm, corresponding to $\mathbf{CH_2}$, **CH** and aromatic protons in polystyrene. The quantitative analysis of copolymer composition was calculated by the ratio of two integrated intensities between aromatic protons (δ = 6.4-7.3 ppm) in PS side chains and methylene protons (δ = 1.35-1.55 ppm) and the number of protons both chemical shifts represent. Figure 2 (a), (b) and (c) indicate 25.6, 38.1 and 43.8 mole % of PS, respectively, in PE-g-PS copolymers. The graft copolymers were analyzed by DSC at a heating rate of 20°C a minute to determine the crystalline melting behavior. Figure 3 shows the DSC curves of three PE-g-PS copolymers containing 17, 33, and 57 weight % of PS. All samples were given the same thermal treatment by heating in a Mettler hot stage at 180°C for 15 minutes

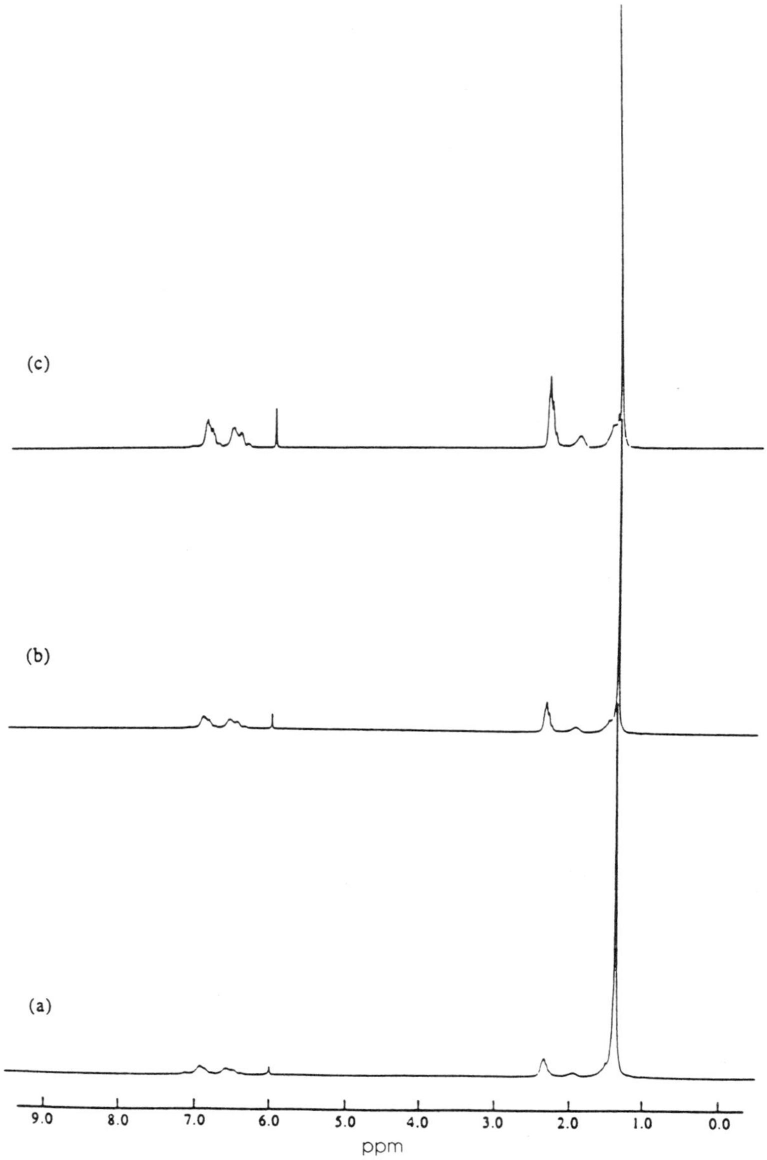

Figure 2. The ^1H NMR spectra of PE-g-PS copolymers, containing (a) 17 (b) 33 and (c) 57 mole % of polystyrene.

before cooling quiescently. Two distinctive crystalline structures are formed in all graft copolymers. It is clear that the high melting peak at about 130 °C is due to the polyethylene segments in the backbone and the low temperature transition at about 90 °C is due to the glass transition temperature of PS segments in the side chains. A slight decrease in the melting point of PE, however, two phases are clearly separated. Table 2 reveals some trends in the graft copolymers' crystalline morphologies. As the weight percent of PS in the graft copolymer increases, so does the Tg. Since the number of growing sites in the backbone is the same, then the number of grafts per chain is constant. Therefore, the increasing PS

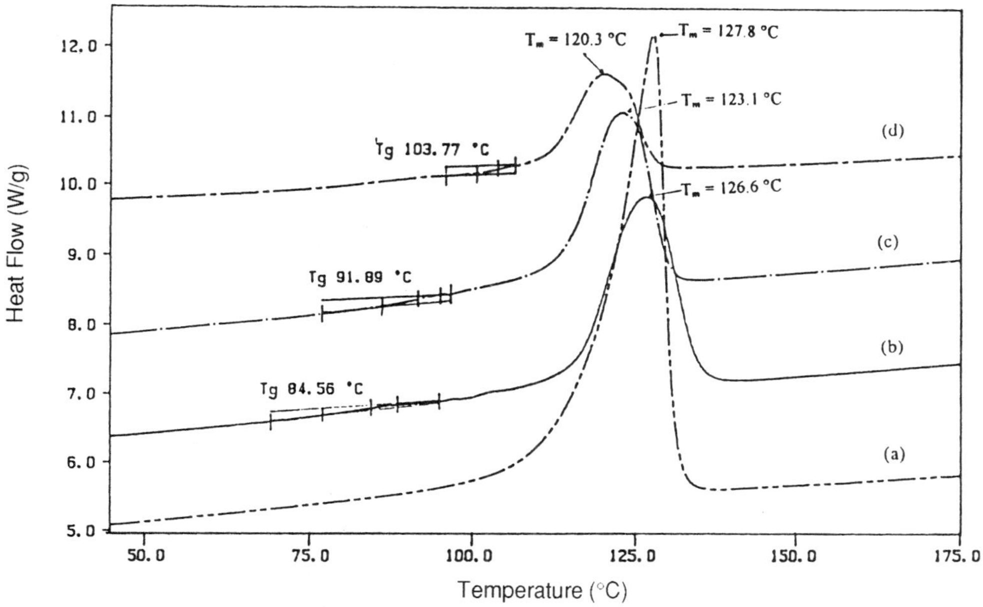

Figure 3. DSC curves comparison between (a) starting poly(ethylene-co-p-methylstyrene) and (b) PE-g-PS copolymers, containing 44 % of polystyrene.

content reflects only increasing PS graft length. Increasing the molecular weight of PS increases its Tg as demonstrated by increasing the PS segments on the graft copolymers. The ethylene segment lengths remain constant, since the amount of graft sites are constant. However, the increasing side chain (graft) length of the PS should cause some change. The PS length may force the distorsion of PE segments and lower in T_m and ΔH.

Table 4. A summary of the anionic graft-from polymerization reactions by using lithiated poly(ethylene-co-p-methylstyrene) as initiator

Lithiated Polymer (g)	Comonomer[1] /(g)	Solvent	Temp. °C	Time hr.	isolated Copolymer (g)	Comonomer in Copolymer (mole%)
1.5	ST/1.9	hexane	25	1	3.3	24.4
1.2	ST/5.9	hexane	25	1	6.8	54.7
1.0	MMA/3.7	THF	0	1.5	1.86	20.0
1.0	MMA/3.4	THF	0	15	2.66	31.8
0.8	MMA/4.0	hexane	25	5	3.08	44.4
0.8	MMA/4.0	hexane	0	5	2.21	33.0
1.0	AN/3.0	hexane	25	16	2.99	51.2
1.0	p-ms/4.0	hexane	25	0.5	5.0	48.7

[1]ST: styrene, MMA: methyl methacrylate, AN: acrylonitrile, p-ms: p-methylstyrene

Table 4 summarizes the reaction conditions and the experimental results. Overall, the experimental results clearly show a new class of PE graft copolymers which can be conveniently prepared by the tranformation of metallocene catalysis to anionic graft-from polymerization.

PE/PS Polymer Blends

It is interesting to study the compatibility of PE-g-PS copolymer in HDPE and PS blends. Polarized optical microscope and the SEM were used to examined the surfaces and bulk morphologies, respectively. Two blends comprised of overall 50/50 weight ratio of PE and PS, one is a simple mixture of 50/50 between HDPE and PS and the other is 45/45/10 weight ratio of HDPE, PS, and PE-g-PS with 50 mole % PS.

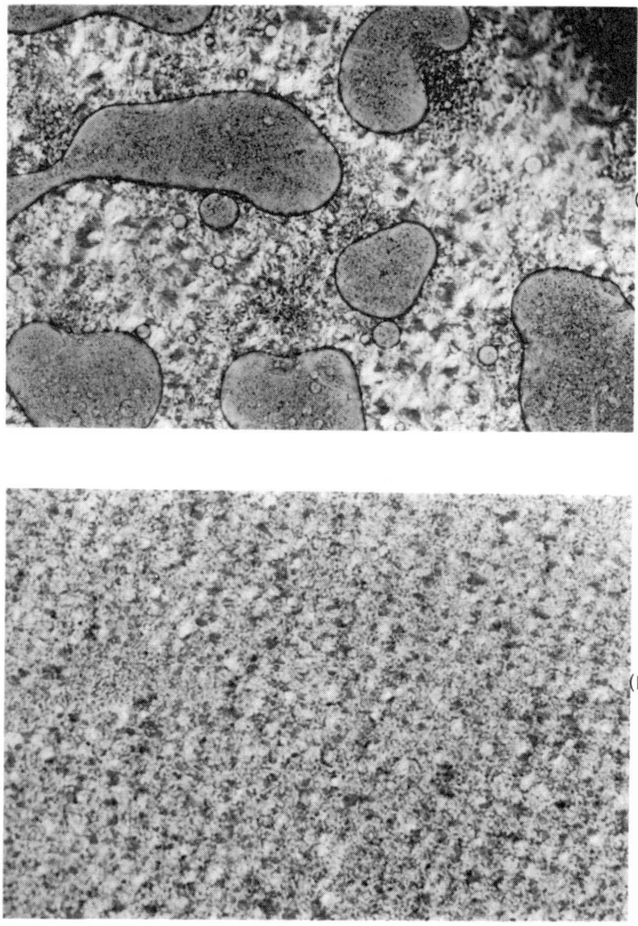

(a)

(b)

Figure 4. Polarized optical micrographs of polymer blends, (a) two homopolymer blend with PE/PS = 50/50 (100x), (b) two homopolymers and PE-g-PS copolymer blend with PE/PE-g-PS/PS = 45/10/45 (100x).

Figure 4 compares the polarized optical micrograph of two blends which were prepared by dissolving polymer mixture in chlorobenzene and cast on a glass slide. Two optical patterns are very different. A gross phase separation in Figure 4 (a) shows the spherulitic PE and the amorphous PS phases. The PS phases vary widely in both size and shape due to the lack of interaction with the PE matrix. On the other hand, the continuous crystalline phase with only small distorted spherulites in Figure 4 (b) shows the compatibilized blend. Basically, the large phase seperated PS domains are now dispersed into the inter-spherulite regions and cannot be resolved by the resolution of the optical microscope. The graft copolymer behaving as a polymeric emulsifier increases the interfacial interaction between the PE crystalline and the PS amorphous regions to reduce the domain sizes.

Figure 5 shows the SEM micrographs, operating with secondary electron imaging, which show the surface topography of cold fractured film edges. The films were cryro-fractured in liquid N_2 to obtain an undistorted view representitve of the bulk material.

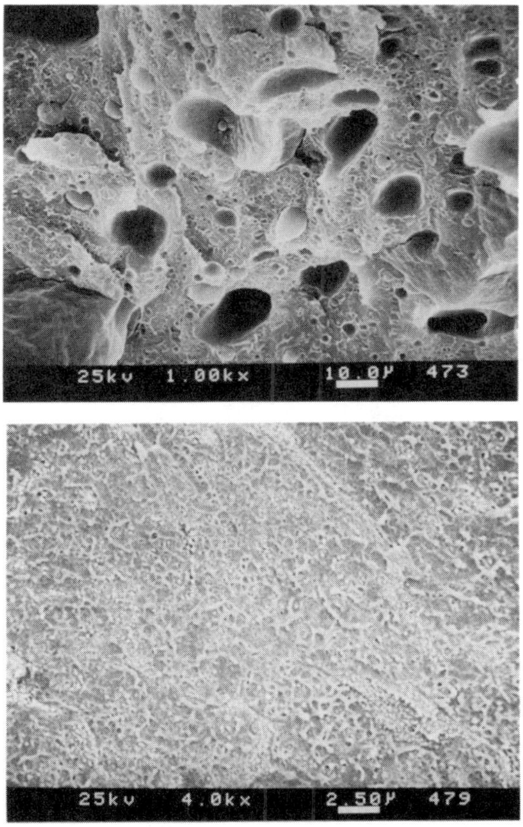

(a)

(b)

Figure 5. SEM micrographs of the cross-section of two polymer blends (a) two homopolymers with PE/PS =50/50 (1,000x) (b) two homopolymers and PE-g-PS copolymer blend with PE/PE-g-PS/PS = 45/10/45 (4,000x).

In the homopolymer blend, the polymers are grossly phase separated as can be seen by the minor component PS which exhibits non-uniform, poorly dispersed domains and voids at the fracture surface as shown in Figure 5 (a). This "ball and socket" topography is indicative of poor interfacial adhesion between the PE and PS domains and represents PS domains that are pulled out of the PE matrix. Such pull out indicates that no stress transfer takes place between phases during fracture. The graft copolymer shows a totally different results in Figure 5 (b). The material exhibits flat mesa-like regions similar to pure PE. No distinct PS phases are observable indicating that fracture occurred through both phases or that the PS phase is too small to be observed. The PE-g-PS is clearly proven to be an effective compatibilizer for PE and PS blends.

EXPERIMENTAL DETAILS

Instrumentation and Materials

All ^{1}H NMR were run on a Bruker AM 300 instrument in either d_6-benzene at room temperature or in d_{10}-*o*-xylene at 130°C. DSC analysis were run on a DuPont Instruments DSC-2910 with Thermal Analyst-2000 controller. In the optical microscopy studies, the samples were observed and photographed with an Olympus BH2 microscope under cross polarizers mounted with a 35 mm C-35A-4 camera and an Olympus AD Systems exposure control unit. The samples were prepared by casting thin fims onto glass microscope slides and covering with a cover slip. The PE based polymers were usually cast from dilute solution of hot xylene or chlorobenzene. Scanning electron microscopy was used to view some of the polymer films with a Topcon International Scientific Intruments ISI-SX-40 using secondary electron imaging. SEM samples were prepared from films cryo-fractured in liquid N_2, and were mounted on an aluminum stub and carbon coated to form a conductive coating.

All O_2 and moisture sensitive manipulations were carried inside of an argon filled Vacuum Atmosphere dry box. HPLC grade hexane, tetrahydrofuran and toluene were deoxygenated by argon sparge before refluxing for 48 hours and then distilling from their respective green or purple sodium anthracide solution under argon. Isopropanol was refluxed in CaH_2 before distilling under argon. All three solvents were stored in the dry-box. Both catalysts,$[C_5Me_4(SiMe_2N^tBu)]TiCl_2$ and $Et(Ind)_2ZrCl_2$, were prepared by the published procedures [20,21]. Methylaluminoxane (MAO) (Ethyl) and n- and s- BuLi (Aldrich) was purchased and used as received.

Copolymerization of Ethylene and p-Methylstyrene

In a typical ethylene copolymerization condition, the comonomer (i.e. p-methylstyrene) was mixed with solvent (toluene or hexane) and methylaluminoxane (MAO) (30 wt% in toluene) needed in a Parr 450 ml stainless autoclave equipped with a mechanical stirrer. The sealed reactor was then saturated with 45 psi ethylene gas at 30 or 53 °C before the adding catalyst solution, $[C_5Me_4(SiMe_2N^tBu)]TiCl_2$ in toluene, to initiate the polymerization. Additional ethylene was fed continuously into the reactor by maintaining a constant pressure about 45 psi during the whole course of the polymerization. After 60 min, the reaction was terminated by adding 100 ml of dilute HCl solution in MeOH. The polymer was isolated by filtering and washed completely with MeOH and dried under vacuum at 50 ^{0}C for 8 hrs.

Lithiation and Silylation Reactions of Poly(ethylene-co-p-methylstyrene)

In an argon filled dry box, 10 g of poly(ethylene-co-p-methylstyrene) with 2.76 mol% of p-methylstyrene (9.08 mmol) was suspended in 100 ml of cyclohexane in a 250 ml air-free flask with a magnetic stirr bar, 21 ml (27.3 mmol) of 1.3 M s-BuLi solution and 4.2 ml (27.3 mmol) of TMEDA were added to the flask, the flask was then brought out of dry box and heated up to 60 ^{0}C for 4 hrs under N_2, the reaction was then cooled down to room temperature and moved back to the dry box. The resulting polymer was filtered and washed completely with cycohexane untill decoloration of filtrate, then was dried under vaccum. A lithiated yellow polymer podwer was obtained. To study the degree of metallation reaction, some (~ 1 g) of the lithiated polymer was suspended in 20 ml of dry THF, 0.5 g of Me_3SiCl was added and stirred at room temperature for 2 hrs. The resulting polymer was then filtered and washed repeatly with THF, methanol, water and methanol, then was dried under vaccum. ^{1}H NMR spectrum shows a strong peak at 0.05 ppm corresponding to the methyl proton next to Si.

PE Graft Copolymers Prepared by Graft-from Reactions

The lithiated polyethylene copolymer was then suspended in hexane or THF solvent. The graft-from reactions were carried out in slurry solution by reacting the lithiated polyethylene copolymer with anionic polymerizable monomers, such as styrene and p-methylstyrene. After certain reaction time, 10 ml of isopropanol was added to terminate the graft-from reaction. The precipitated polymer was filtered and then subjected to fractionation. Good solvents for backbone and side chain polymers were used during the fractionization, using a Soxhlet apparatus under N_2 for 24 hours. The soluble fractions were isolated by vacuum-removal of solvent. Usually, the total soluble fractions were less than 5 % of the product. The major insoluble fraction was PE graft copolymer, which was completely soluble in xylene or trichlorobenzene at elevated temperatures.

Polymer Blending

All the blends were prepared in solution to obtain molecular level mixing. The PE was first dissolved in refluxing chlorobenzene inhibited with BHT and kept under a N_2 blanket to prevent oxidation. The PE-g-PS and PS were then slowly added to the hot solution. After the polymer mixture had formed a clear, homogeneous solution, the blend was precipitated into cold hexane. The blend was dried under vacuum before being melt pressed to form a film.

CONCLUSION

The major goal of this research is the development of the new chemical routes to prepare polyolefin graft copolymers, for improving the compatibility of polyolefin in polymer blends. The combination of the metallocene catalyst and reactive comonomer approach, i.e. borane containing α-olefin and p-methystyrene, allows a smooth transformation process from coordination olefin polymerization to living free radical and anionic polymerization to produce graft copolymers with well-defined polymer side chains, without any significant amount of homopolymers. In the bulk, the graft copolymers were shown to be good compatibilizers for polyolefin blends, as evidenced by the fine dispersion of domains and the strong interfacial interactions.

ACKNOWLEDGEMENT

Authors would like to thank the Polymer Program of the National Science Foundation for financial support.

REFERENCES

1. M. D. Baijal, "*Plastics Polymer Science and Technology*", John Wily & Sons, New York, (1982).
2. C. Pinazzi, P. Guillaume and D. Reyx, *J. Eur. Polym..* 13:711 (1977).
3. T. C. Chung, M. Raate, E. Berluche and D. N. Schulz,*Macromolecules* 21:1903 (1988).
4. T. C. Chung, *J. Polym. Sci., Polym. Chem. Ed.* 27:3251 (1989).
5. G. Riess, J. Periard, A. Bonderet, *Colloidal and Morphological Behavior of Block and Graft Copolymers*, Plenum: New York, (1971).
6. B. Epstein, U. S. Patents 4,174,358, (1979).
7. G. Natta, E. Beati, and F. Severine,*J. Polymer Sci.* 34:548 (1959).
8. B. Ranby and F. Guo, *Polymer Preprints* 31:446 (1990).
9. T. C. Chung and D. Rhubright, *Macromolecules* 26:3019 (1993).
10. T. C. Chung, H. L. Lu and C. L. Li, *Polymer International* 37:197, (1995).
11. R. L. Bernard, MS Thesis, Penn State University, (1995).
12. T. C. Chung, D. Rhubright and G. J. Jiang, U. S. Patent 5,286,800 (1994).
13. T.C. Chung, G. J. Jiang and D. Rhubright, U. S. Patents 5,401,805 (1995).
14. (a) R. F. Jordan, *J. Chem. Edu.* 65:285 (1988). (b) X. Yang, C. L. Stern, T. J. Marks, *J. Am. Chem. Soc.* 116:10015 (1994).

15. (a) A. Grassi, P. Longo, A. Proto, A. Zambelli, *Macromolecules* 22:104 (1989). (b) K. Soga, H. Nakatani, T. Monoi, *Macromolecules* 23:953 (1990).
16. J. F. Harris, W. H. Sharkey, U.S. Patent 4,145,190 (1979).
17. T. C. Chung and H. L. Lu, U. S. Patent 5,543,484 (1996).
18. F. G. Sernetz, R. Mulhaupt and R. M. Waymouth, *Macromol. Chem. Phys.* 197:1071 (1996).
19. J. C. Randall, *J. of Polym. Sci., Polym. Phy. Ed.* 13:889 (1975).
20. I. M. Lee, W. J. Gauthier, J. M. Ball, B. Lyengar and S. Collins, *Organometallics* 11:2115 (1992).
21. J. C. Stevens, F. J. Timmers, J. R. Wilson, G. F. Schmidt, P. N. Nickias, R. K. Rosen, G. W. Knight and S. Y. Lai, Eur. Pat. Appl. 416,815 A2, 1991.

CHEMICAL MODIFICATION OF ACRYLAMIDE POLYMERS IN AQUEOUS SOLUTION

D. W. Fong and D. J. Kowalski

Nalco Chemical Co.
Naperville, IL 60563

INTRODUCTION

The world market for synthetic water-soluble polymers was about 950,000 metric tons in 1994.[1] Among them, acrylamide copolymers are used extensively in water treatment as flocculants, coagulants and dispersants. The activities of these polymers are closely related to their molecular weights and structures, especially the nature of the pending groups. One can manipulate the polymerization process to achieve the desired molecular weights. Low molecular weight polymers are usually made in solution with chain transfer reagents. Inverse emulsion polymerization and gel polymerization are commonly used to produce high molecular weight polymers. The desired functionalities of the copolymers can be introduced by using proper monomers or by post-modifications of the acrylamide polymers. Copolymerization of acrylamide with the desired comonomer is usually the method of choice for commercial production. For example, copolymerization of acrylamide and 2-(N-acrylamido)methylpropanesulfonic acid (AMPS) salt is used to produce both high molecular weight flocculants and low molecular weight dispersants. Acrylamide and 2-(dimethylamino)ethyl acrylate methochloride (DMAEM. MCQ) copolymerize to produce high molecular weight flocculants. However, this method is sometimes restricted by the availability of the desired comonomers. The solubility, hydrolytic stability, and the reactivity of the comonomer can also cause processing difficulties. In some cases, post-modification of acrylamide copolymers can be a good alternative for producing the desired polymers. The aminomethylation of poly(acrylamide) (Mannich Reaction) to produce cationic coagulants and flocculants,[2] the hydrolysis of poly(acrylonitrile) to poly(acrylic acid) as dispersants,[3] and the sulfonation of styrene-maleic anhydride copolymers to produce sulfonated dispersants[4] are commonly practiced in the industrial water treatment industry. In this work, the modification of acrylamide copolymers with some primary amines was studied.

EXPERIMENTAL

Poly(acrylamide) and its acrylic acid copolymers were obtained from polymerization of the monomers in aqueous solution using ammonium persulfate/sodium bisulfite initiator following the procedure reported by Fong and Kowalski.[5] Depending on the amount of initiator used, the average molecular weights of the polymers were between 6,000 to 50,000 as determined by gel permeation chromatography (GPC). Taurine (2-aminoethanesulfonic acid) and sodium formaldehyde bisulfite were purchased from Aldrich and used without purification.

The post-modification reactions of acrylamide polymers were run as follows.[6] A solution of poly(acrylamide[75mol%]-acrylic acid) (150 g, 27.5% in water) and sodium formaldehyde bisulfite (15.5 g), pH 4.3, was heated to 150° C and maintained at that temperature for four hours in a 300 ml Parr reactor which was equipped with a mechanical stirrer and a thermocouple. The pH of the resulting polymer solution at room temperature was 5.9.

A solution of poly(acrylamide[65mol%]-acrylic acid) (100 g. 30% in water) and taurine (29.1 g), pH 5.4, was heated to 163° C and maintained at that temperature for 4.5 hours in a 300 ml Parr reactor which was equipped with a mechanical stirrer and a thermocouple. The pH of the resulting polymer solution was 6.5.

Transmission IR spectra were obtained on a Beckman 4260 infrared spectrometer. The aqueous samples were cast on silver chloride plates and the solvent was evaporated using a nitrogen stream without heating.

Carbon-13 NMR spectra were obtained using a JEOL FX90Q operating at 22.5 MHz for carbon. Spectra were obtained using 75° pulses and a 4 sec. delay between pulses with complete proton decoupling. The samples were analyzed in 10 mm tubes.

Ionic charges of the polymers were determined by photometric colloid titrations in some instances. A known amount of poly(diallyldimethylammonium chloride) was added to the polymer solution at a pH of 2.5. The excess poly(diallyldimethylammonium chloride) was titrated by poly(vinylsulfate) using the adsorption indicator methylene blue. The end point was detected by the photometric detector as the color of the solution changes from blue to violet. For anionic copolymers the colloid titration was conducted at pH values of 2.5 and 10.0 to determine the extent of modification.

RESULTS AND DISCUSSION

The reaction of acrylamide copolymers and taurine was studied at temperatures between 125° and 200° C, reaction time 2-7 hours, and taurine charge 10-100 mol% based on polymer. The substituted amide formation was determined by NMR and colloid titration. The C-13 NMR of the product exhibits carbonyls consistent with the formation of a secondary amide. The spectrum also exhibits two new methylene signals for the incorporated taurine at chemical shifts slightly different from the starting taurine. Additionally, the chemical shifts for the signals of taurine are pH dependent, whereas little change in chemical shift is observed for the signals of the incorporated taurine. The presence of sulfonate incorporated into the polymer was detected and quantitatively determined by colloid titration at pH 2.5.

The conversions of taurine to secondary amide (sulfoethylamide) under various reaction conditions are summarized in Table I. At 10 mol% charge, the conversion of taurine to secondary amide increased with increasing reaction time and temperature. At 100 mol% charge, the conversion was only about 50%. Under similar reaction conditions, taurine reacted more favorably with copolymers than with homopolymers. In general, the final sulfonate content increases with increasing initial taurine charge and increasing reaction temperature. The percent conversion generally increases with increasing reaction temperature and reaction time, and with decreasing initial taurine charge. Under most favorable conditions the conversions were unexpectedly high, considering that amides would hydrolyze under the same reaction conditions.

Both aliphatic and aromatic amines have been investigated in this reaction with similar results. This confirms that the reaction is a general reaction with amines in aqueous medium. The conversions of aromatic amines to secondary amides are lower than that of the aliphatic amines. This is because aromatic amines are generally weaker bases and poorer nucleophiles.

Acrylamide polymers in aqueous solution undergo thermal hydrolysis and cyclic imide formation. Acrylate, acrylamide and cyclic imide functional groups were detected when a poly(acrylamide) is heated at 150°C in water. The formation of intramolecular imide has been reported in literature. Moradi-Araghi, Hsieh and Westerman reported the formation of cyclic imide in acid, neutral and slightly basic media at 90° C.[7] In acidic media, imide formation is favored. In neutral and basic media, both hydrolysis to acrylate and imide formation do occur, but hydrolysis is the dominant reaction. We speculate the high conversion of amine to amide is the result of transamidation, amidation and the nucleophilic addition of the amine to the glutarimide intermediate (Reaction 1).

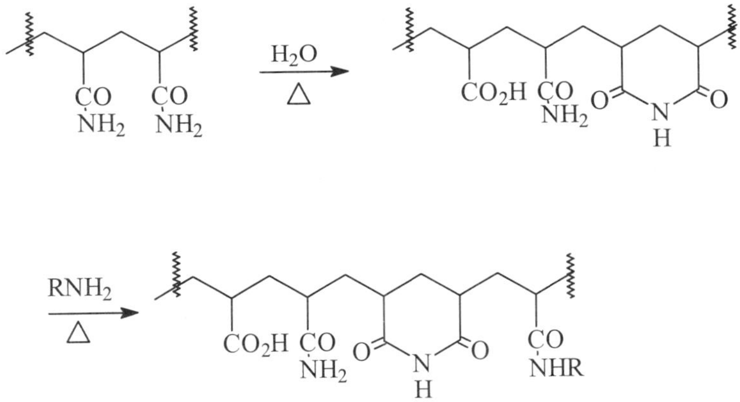

(1)

Samples taken from the acrylamide copolymer and taurine reaction mixture were analyzed. A typical C-13 NMR spectrum is shown in Figure 1. It exhibits an acrylate multiplet at 184 ppm, an acrylamide multiplet at 180 ppm and a multiplet at about 178 ppm for both secondary amide and imide carbonyls. The methylene from the sulfoethylamide attached to sulfonate occurs at 50 ppm while the methylene attached to the amide nitrogen overlaps with the methylene of the backbone but is observable at 36 ppm. The methylene of taurine attached to the sulfonate occurs at 49 ppm while the other methylene occurs with the backbone methylene signal. The backbone methine signal for the acrylate portion of the polymer occurs at 46 ppm while the backbone methine signal for the amide occurs at 44 ppm.

The reaction of polyacrylamide with formaldehyde and sodium bisulfite or sodium formaldehyde bisulfite at temperatures below 100° C to form amidomethylsulfonate was studied by Schiller and Suen.[8] They reported sulfomethylation readily occurred at pH ~12 and temperature at 70°-75° C. The conclusion was mostly based on the results of redox titrations for residual bisulfite and sulfite. Assuming all residual bisulfite and sulfite are titratable and the formation of sodium hydroxymethane sulfonate from formaldehyde and bisulfite is substantially complete at low pH values, only the bisulfite that is in excess of formaldehyde can be titrated. At high pH values, the formation of sodium hydroxymethane

Table I. Sulfoethylation of acrylamide polymers

Poly(AcAm/AA) mole ratio	Temp. °C	Reaction time hours	Taurine charge mole % based on polymer	% Conversion based on taurine[1]
100/0	125	2	10	13
100/0	200	2	10	75
65/35	163	2	10	73
65/35	125	7	10	100
30/70	163	7	10	100
0/100	160	4.5	50	21
100/0	163	7	55	73
65/35	163	4.5	55	84
100/0	200	2	100	49
30/70	200	4.5	100	50

[1] Determined by colloid titration.

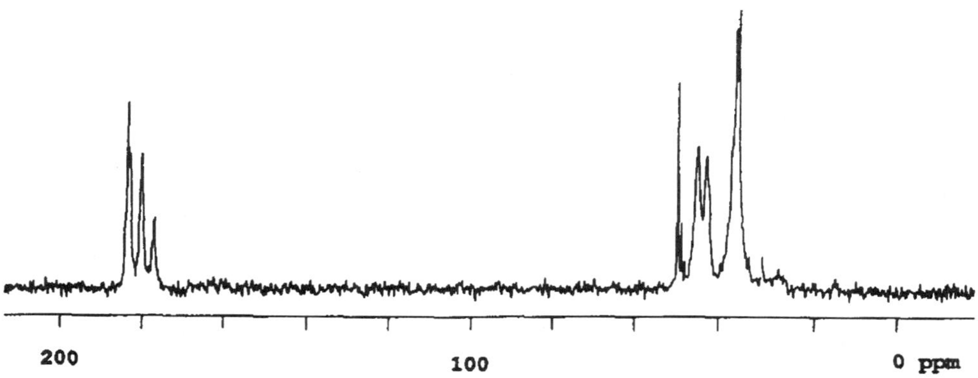

Figure 1. C-13 NMR of taurine modified poly(acrylamide-acrylate)

sulfonate is not favorable, and the bisulfite and sulfite that are not combined with poly(acrylamide) can be titrated. With these assumptions the degree of sulfomethylation was calculated.

Bakalik and Kowalski investigated the sulfomethylation reaction in some detail using C-13 NMR.[9] They found there was no detectable amount of amidomethylsulfonate under the condition described by Schiller and Suen. The major products are partially hydrolyzed poly(acrylamide) and the mono, bis, and tris sulfomethylamines. Since the sulfomethylamines are not titratable by the redox titration, the iodometric titration results are not a good indicator for the formation of amidomethylsulfonate product.

However, the reaction of acrylamide copolymer with sodium formaldehyde bisulfite at pH 4-7 and at temperatures above 125° gave the expected product, amidomethylsulfonate (Table II). A typical C-13 NMR spectrum is shown in Figure 2. A typical IR spectrum is shown in Figure 3. The presence of a secondary amide is shown by the absorption band at 1550 cm^{-1} after the sample was acidified to pH 1 to remove carboxylate salt interference (Figure 4). The amidomethylsulfonate group is hydrolytically stable both at high and low pH values at room temperature. The sulfonic acid groups on the polymer are strong acid groups which can be quantitatively determined by colloid titration at pH 2.5. We attribute the amidomethylsulfonate formation to the reaction of acrylamide copolymer with aminomethylsulfonate which was formed from sodium formaldehyde bisulfite and ammonia generated from amide hydrolysis (Reaction 2).

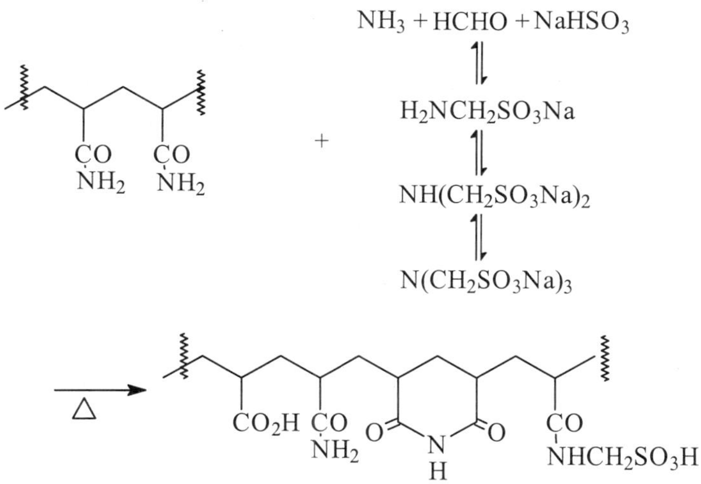

(2)

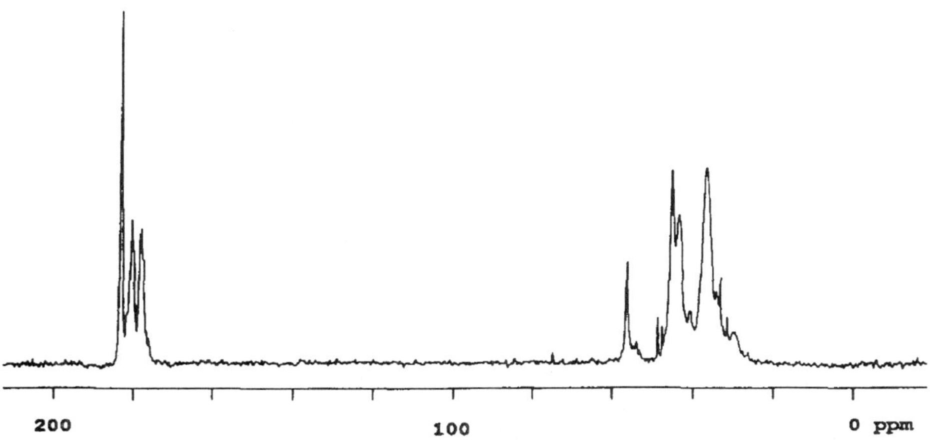

Figure 2. C-13 NMR of sulfomethylated poly(acrylamide)

The data in Table II show that under the same reaction condition as described in Table I, the conversion of aminomethylsulfonate to the amidomethylsulfonate was the same as that of the taurine.

Table II. Sulfomethylation of acrylamide polymers

Poly(AcAm/AA) mole ratio	Temp. °C	Reaction time hours	SFB charge mole % based on polymer[1]	% Conversion based on SFB
100/0	120	7	15	3[2]
75/25	150	4	20	>95[3]
75/25	150	4	60	42[2]

[1] SFB - Sodium formaldehyde bisulfite.
[2] Determined by colloid titration.
[3] C-13 NMR estimate.

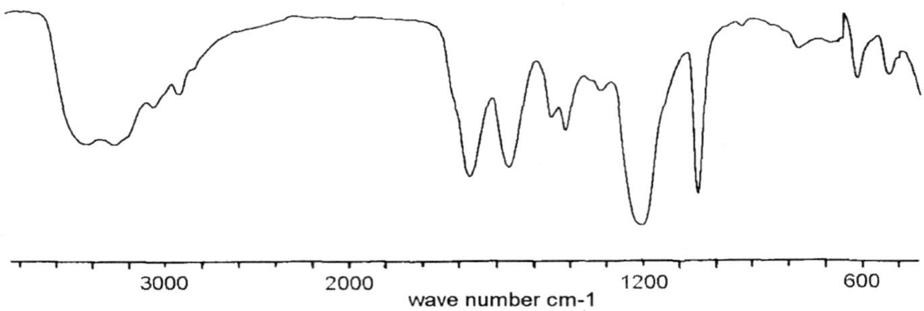

Figure 3. IR spectrum of sulfomethylated poly(acrylamide) at pH 6.

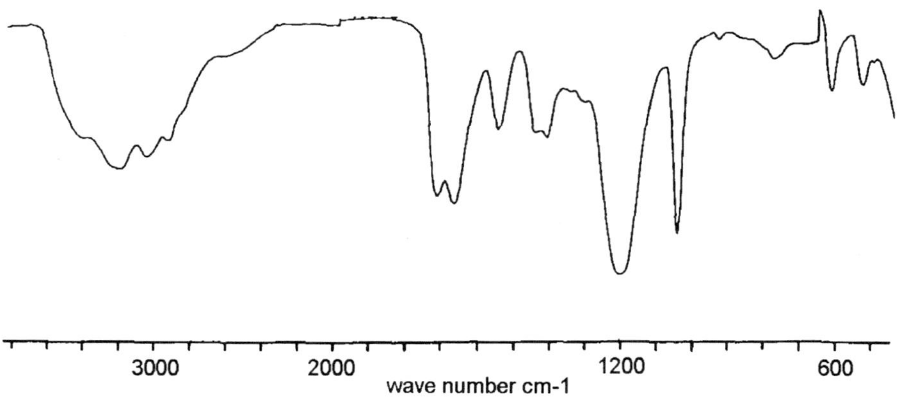

Figure 4. IR spectrum of sulfomethylated poly(acrylamide) at pH 1

CONCLUSIONS

It has been demonstrated that, at temperatures above 125° C, acrylamide copolymers react with amines in aqueous solution to form the corresponding substituted amides in high yields. The process is versatile and a wide variety of N-substituted acrylamide copolymers can be prepared by this method.

REFERENCES

1. J. C. Goin, M. Jackel and Y.Ishikawa, *Water-Soluble Polymers* in "Specialty Chemicals," SRI International, CA, Vol. 14, December 1994, p. 11.
2. R. F. Krebbs, P. J. Marek and K. G. Phillips, U.S. Patent 4,405,728 (1983).
3. W. P. Hettinger, U.S. Patent 3,492,240 (1970).
4. W. B. Chiao and D. K. Ray-Chaudhuri, U.S. Patent 4,450,261 (1984).
5. D. W. Fong and D. J. Kowalski, *J. Polym. Sci.: Part A: Polym. Chem.,* **31**, 1624-1627 (1993).
6. D. W. Fong, U.S. Patent 4,703,092 (1987); D. W. Fong and D. J. Kowalski, U.S. Patent 5,120,797 (1992).
7. A. Moradi-Araghi, E. T. Hsieh and I. J. Westerman, *Water-Soluble Polym. Pet. Recovery*, [Proc. Natl. Meet. ACS] (1988), Editor(s): G. A. Stahl, D. N. Schulz, Plenum, New York, N. Y.**;** J. S. Shepitka, C. E. Case, L. G. Donaruma, M. J. Hatch, N. H. Kilmer, G. D. Khune, F. D. Martin and K. V. Wilson, *J. Appl. Polym. Sci.,* **28**(12), 3611 (1983).
8. A. M. Schiller and T. J. Suen, *Indust. Eng. Chem.,* **48**, 2132 (1956).
9. D. P. Bakalik and D. J. Kowalski, *J. Polym. Sci.: Part A: Polym. Chem.,* **25**, 433-436 (1987).

SYNTHESIS AND CHARACTERIZATION OF POLY(MAGNESIUM ACRYLATE) AND POLY(ZIRCONYL ACRYLATE) TOWARDS THE FORMATION OF MAGNESIUM PARTIALLY STABILIZED ZIRCONIA CERAMICS

Xinhua Xu and Charles E. Carraher, Jr.
Florida Atlantic University
Boca Raton, FL 33431 and
Florida Center for Environmental Studies
NorthCorp Center
Palm Beach Gardens, FL 33410

INTRODUCTION

Ceramics are no longer only traditional materials such as porcelain and earthenware. Today, the term "ceramics" covers a wide range of inorganic materials that includes not only traditional materials but also some glass, cement and concrete, and special or advanced ceramics. Most advanced ceramics, such as zirconia ceramics, can not be successfully manufactured using traditional ceramic processes because they require high purity materials and more carefully controlled compositions. Chemical methods have been used in attempts to achieve modern ceramic materials. The present approach is a new approach and is aimed at providing a method of controlling mixing on a molecular level.

Mg-PSZ represents magnesia, MgO, partially stabilized zirconia, ZrO_2, ceramics. These ceramics are strong and tough structural materials. The main characteristics of Mg-PSZ ceramics are

*high bending strength (2 to 3 times that of corundum or alumina ceramics)

*high fracture toughness (which means it is not as brittle as traditional ceramics; it is 3 to 5 times better than corundum ceramics)

*high resistance to wear and corrosion and

*high density (5.8 gram/cc).

All of these advantages make Mg-PSZ ceramics applicable to many industrial fields such as in the construction of oil extraction pump ball valves in the petroleum industry, milling balls in the materials industry, shear blades in the textile industry, plungers in the food and drink industry, etc.

Structural Properties of Mg-PSZ

Zirconia exists in three solid phases-cubic, tetragonal and monoclinic. It undergoes the following transformations

MONOCLINIC<--1170 C-->TETRAGONAL<--2370 C-->CUBIC<--2680 C-->LIQUID

The transformation between monoclinic and tetragonal involves a large and abrupt volume change. Thus, zirconia ceramics undergo a substantial expansion on cooling through the TETRAGONAL to MONOCLINIC transformation, leading to a crumbling of the ceramic.

As an attempt to solve this problem, zirconia is "stabilized" in the cubic phase by alloying it with an appropriate amount of di- or tri-valent oxide of cubic symmetry such as CaO, MgO or Y_2O_3. This results in a lowering of the temperature for the two lowest temperature transitions. These alloys are called partially stabilized zirconia, PSZ and they are a mixture of cubic and monoclinic or tetragonal phases and fully stabilized zirconia (all cubic phase) depending upon the concentration of the "dopant" or added metal oxide.

The incorporation of alloying metal oxides not only decreases the transition temperatures, but also decreases the thermal expansion coefficient of the two-phase material and the accompanying volume change associated with the monoclinic-tetragonal phase change. The thermal expansion coefficient of PSZ is lower than that of pure zirconia and also that of fully stabilized zirconia. This contributes to a higher strength and greater toughness of the PSZ in comparison to that of the cubic (fully stabilized) and monoclinic (pure) zirconia.

The microstructure of PSZ generally consists of a cubic zirconia solid solution as the major phase with a monoclinic or tetragonal zirconia solid solution as the minor precipitate phase. The second phase may exist at grain boundaries either from the sintering process or by precipitation during post-sintering heat treatment or during cooling, or within the cubic matrix grains. If the tetragonal grain size is less than 0.2 mm, it will retain the tetragonal symmetry on cooling. Larger particles (> 0.2 mm) will spontaneously convert to the monoclinic form which involves a volume increase leading to micro-cracking. Because of the large numbers of cracks, they propagate only quasi-statically allowing the sample to retain much of its strength.

When force is applied to a material, the effects of the stress will most likely appear at the tip of the cracks. These stresses, near the crack tips, will cause the particles to transform to the monoclinic symmetry by making the particles lose their coherency. Additional stress is then required for crack extension, i.e. application of additional stress can inhibit crack propagation thereby strengthening and toughening the PSZ.

Mg-PSZ Processing

General chemical formulas of Mg-PSZ ceramic materials are seemingly very simple. Following is one such formulation.

8-10% (by mole) magnesium oxide
90-92% (by mole) zirconia

The basic process steps are also simple and involve
MIXING<--->SHAPING<--->SINTERING.

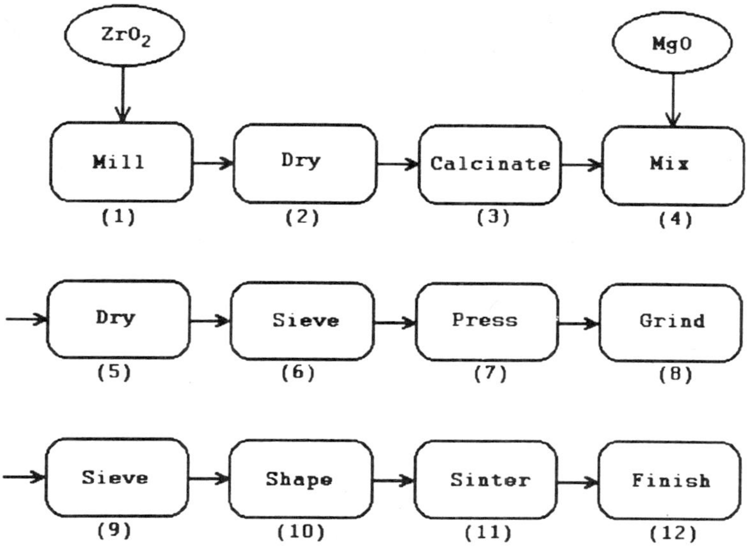

Figure 1. Flow chart used for Mg-PSZ processing.

The most important step in the process is mixing which nvolves

*the proportion of materials to be mixed
*the sizes of the particles and
*how thoroughly they are mixed.

classical processing outline for Mg-PSZ ceramics is given in igure 1.

All of the steps preceding "Shape", step 10, Figure 1, are ntended to make magnesium oxide and zirconia particles smaller and o mix them evenly. Figure 2 contains an illustration of what a lassical "well mixed" pre-ceramic mixture might look like. Figures and 4 are represenations of the crystal structures of magnesium xide and zirconium (IV) oxide.

This processing sequence requires relatively low production ost. Even so, the products often do not meet high technical equirements. The main problem is poor uniformity of the finished roducts due to the following disadvantages of this processing rocedure.

*While the ball mills provide mechanical porphyrization of the agnesium oxide and zirconia particles, it is difficult to get articles with a smaller size than 1 mm. This prohibits sintering t lower temperatures.

*Raw materials are only mixed in a mechanical way so that the ixture of magnesium oxide and zirconia occurs only at the particle evel. This increases the difficulty for the diffusion of agnesium oxide into the crystalline zirconia to stabilize it by orming a magnesium oxide-zirconia solid solution.

*Unevenly mixed mixtures may be obtained due to the ignificantly different powder densities between magnesium oxide nd zirconia.

One way to decrease these problems is to decrease the size of he particles. Unfortunately, the traditional processing sequence s limited with respect to achieving mixing on a molecular scale.

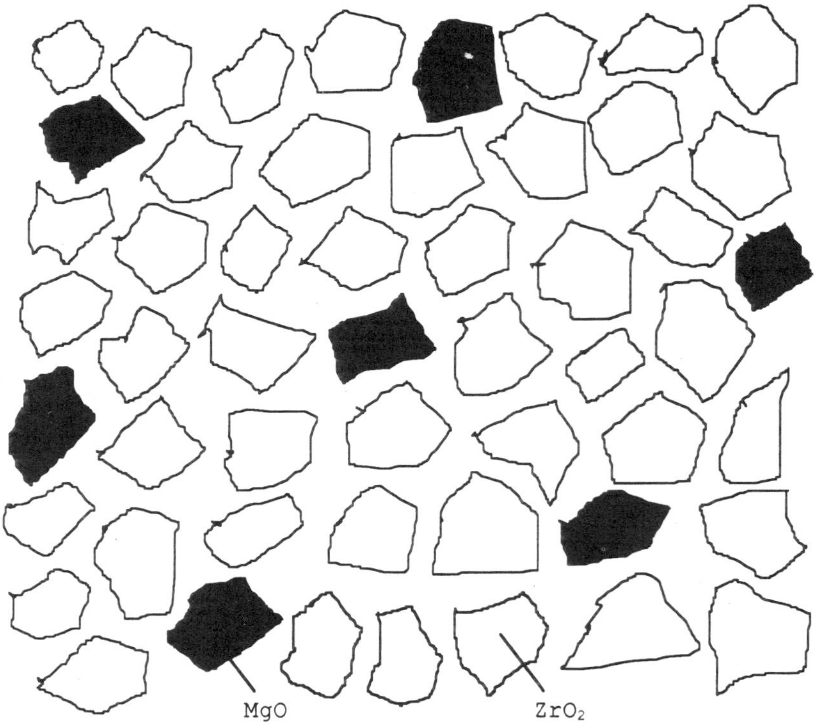

Figure 2. Ball Milled Mixture of MgO and ZrO_2.

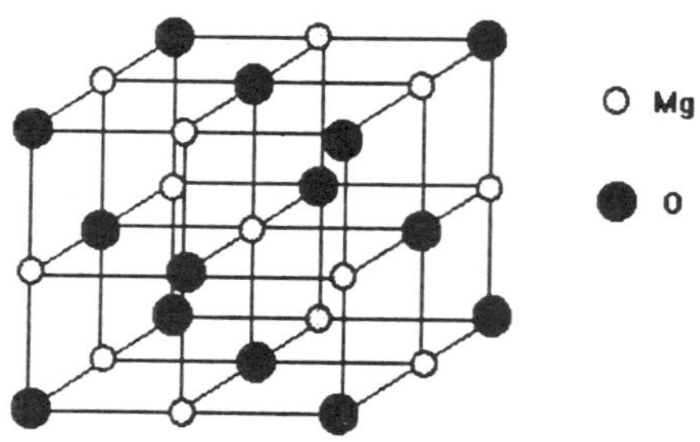

Figure 3. Crystal Structure of MgO.

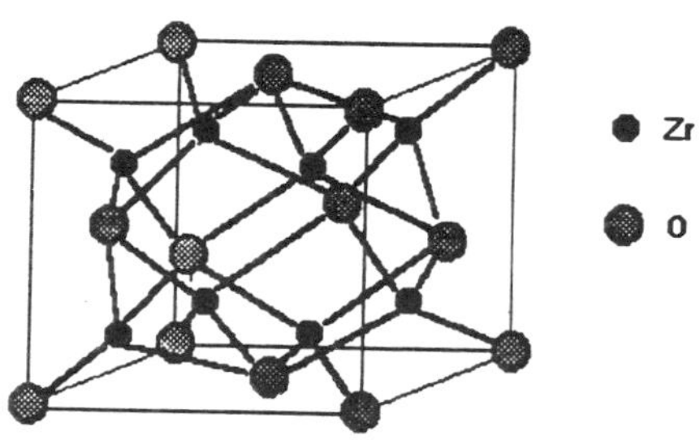

Figure 4. Crystal structure of cubic phase ZrO_2.

Advanced Methods Towards Making Pre-Ceramic Powders

The inability of the current processing sequence to achiev mixing on a molecular level prohibits the production of cerami materials with properties approaching potentially possible limits. Several chemical methods have been employed in an attempt to produce uniform ceramic powders. These include the following:

(A) Freeze dry chemical method in which the solvent is frozrn and then it is removed by sublimation giving an uniform ceramic powder(1).

(B) Sol-Gel Synthesis (2-4) which is a processing in a liquid medium to obtain a solid that does not settle under gravity or does not precipitate.

(C) Alkoxide method (5-7) in which an extremely facile hydrolysis of the metal alkoxides readily leads to the formation of the corresponding metal hydroxides.

(D) Co-precipitation method (8) in which two or more metal ions that precipitate under similar conditions are precipitated simultaneously. While this method can produce very fine particles of magnesium oxide and zirconia, it is difficult to control the amount of magnesium and zirconium that is co-precipitated giving an uneven mixture of magnesium oxide and zirconia. This is because the zirconium (IV) hydroxide precipitates at a pH of about 8.5 while magnesium hydroxide precipitates at a pH of about 11. Further, some of the magnesium hydroxide is rinsed away because it has a small solubility in water at lower pH values resulting in a loss of magnesium.

Present Approach

In this study, magnesium and zirconyl ions are grafted onto a polymer chain. A representative structure is given in Figure 5. This metal-containing polymer is dried and heated to burn away the organic portion leaving behind a fine mixture of magnesium oxide and zirconia.

The nature of the polymer chosen to be grafted onto is important. It should

*be stable under normal conditions allowing the processing steps to be carried out under normal conditions.

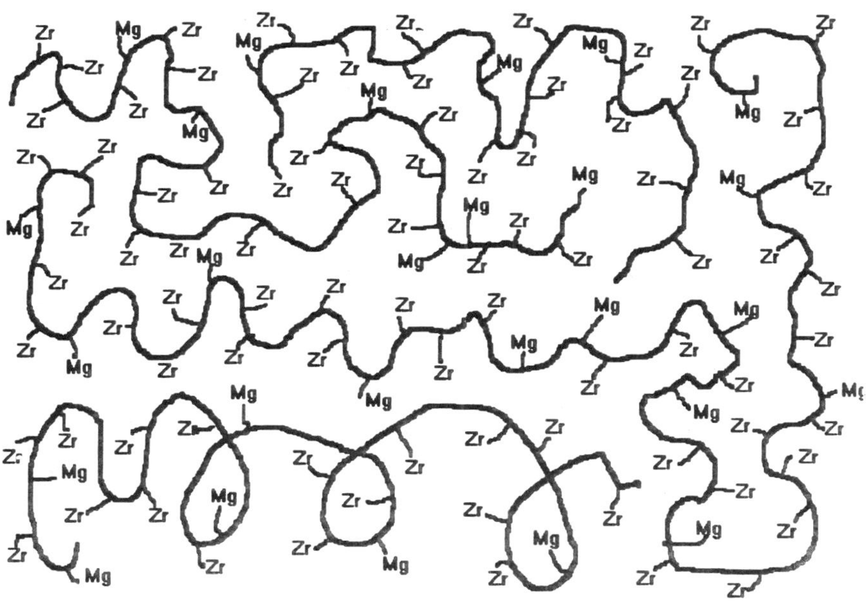

Figure 5. Metal-containing Polymer.

*be soluble in water allowing the ready reaction of water soluble salts of magnesium and zirconium. Further, in industry i is advantageous to employ water solutions rather than using a organic solvent, thus eliminating the need of disposing an protecting against the organic liquid.

*be free of carbon-carbon double or triple bonds. The polyme salt will be heated to get rid of the organic material. Double o triple bonded compounds may form complicated cyclic residues whic will remain within the ceramic after heating.

*contain some oxygen that will be connected to the meta atoms. This metal-containing polymer will be heated to remove al carbon and hydrogen. But it will also contain metal oxides. ready source of oxygen is thus advantageous.

*offer minimal interference with ceramic processing if ther is any trace of un-burned residue. (Poly(acrylic acid), PAA, i already used in the present processing to form Mg-PSZ ceramics an is not believed to cause a problem.)

*be commercially available and low cost so that it will no add much to the cost of the ceramic material.

Poly(acrylic acid) meets the above qualifications.

EXPERIMENTAL

Raw Materials

The following materials were used as received without furthe treatment.

Poly(acrylic acid)-Molecular weight 80,000; CAS RN 9003 01-04; Polysciences, Warrington, PA.

Magnesium sulfate-CAS RN 7487-88-9; Matheson, Coleman ar Bell, Norwood, OH.

Zirconyl chloride-CAS RN 13520-92-8; Aldrich, Milwaukee WI.

Physical Characterization

Poly(acrylic acid), PAA, was titrated to determine the amoun of PAA.

FT-Infrared spectroscopy was accomplished employing KB pellets, using a Galaxy Series FT-IR 4020, Mattson Inst., Madison WI. Thermal analysis was done employing a Dupont Therma Instrument Model 990 employing a heating rate of 20 ^{o}C/min and gas flow rate of 0.1 l/min.

Elemental analysis was performed by Galbrith Labs., Knoxville TN. Mass spectrometry was carried out by the Midwest Center fo Mass Spectroscopy, University of Nebraska, Lincoln, NB. Sample were inserted in a glass ampule using a direct insertion prob (DIP-EI-MS). A Kratos MS-50 Mass Spectrometer was used. Spectr were recorded for samples ballistically heated to 450 C.

Synthesis

Known amounts of sodium hydroxide, sufficient to neutraliz the PAA, were added to aqueous solutions of PAA forming poly(sodiu acrylate), PNaA. Aqueous solutions containing magnesium sulfate c zirconyl chloride are then added to the aqueous solution of PNaA Rapid mechanical stirring is employed to insure good mixing Stirring is continued for ten minutes. The precipitate is remove by vacuum filtration. The precipitate is washed five times wit water. The precipitate is then placed in a glass petri dish ar allowed to dry. The dry solid is ground to a fine powder for use for additional analyses.

In other instances the amount of sodium hydroxide employed in the reaction sequence varied from zero to large excesses.

RESULTS AND DISCUSSION

The following focuses on the structural characterization of the pre-ceramic polymers.

Characterization of Poly(acrylic Acid)

The PAA employed in the present study is sold as a 25% aqueous solution. A given amount was removed and dried. The solid content was 27%.

Thermo-degradation, as measured using TGA, is atmospheric dependent. In dry air, PAA begins to lose weight at 75 C with gradual loss occurring until 250 C. This is followed by a more rapid weight loss to 510 C eventually resulting in a final weight retention of 4% at 850 C. In nitrogen, PAA begins losing weight at 170 C with rapid loss occurring between 270 and 310 C with continual loss of weight with a 3% residue at 850 C.

Mass spectral results are consistent with ion fragmentation of the backbone occurring with the most abundant ion fragments corresponding to the general formulas C_xH_y and $C_xH_y(COOH)_z$. Within the infrared region there is a broad peak at 3448 (all IR values are given in 1/cm) assigned to the O-H stretching. The C=O stretch occurs at 1714 and the C-O stretch at about 1244.

Characterization of Poly(magnesium acrylate)

The products of reaction between poly(sodium acrylate) and magnesium sulfate are believed to contain unreacted units and a variety of unreacted and mono and di-reacted products including both bridged and non-bridges structures. Possible bridged structural units include the following.

Structure <u>1</u>

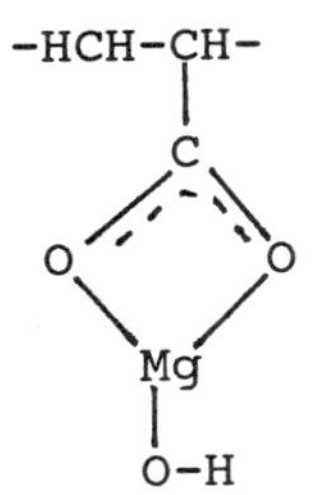

and

Structure <u>2</u>

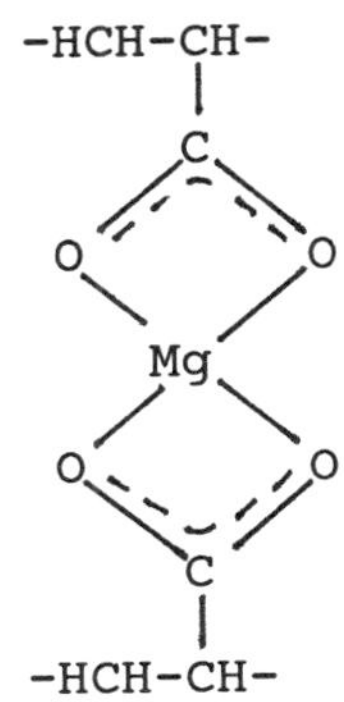

When excess sodium hydroxide is used, formation of magnesium hydroxide occurs as follows.

$$MgSO_4 + NaOH \text{ ---> } Mg(OH)_2 + H_2O$$

Extensive solubility studies of the product and magnesium hydroxide were carried out to allow conditions, procedures, to be developed that would eliminate any magnesium hydroxide that was formed. Tests show that magnesium is readily dissolved in dilute hydrochloric acid (0.4 M) and moderate strength acetic acid (4 M). While PMgA is not soluble in 0.4 M HCl, magnesium hydroxide is. PMgA is soluble in higher concentrations of HCl (1.0 M). Solubility of magnesium hydroxide and PMgA in various ammonium salts was also studied. In general, PMgA is not readily soluble in the ammonium salts tested, but while slow, it is soluble in ammonium citrate. This result may be useful for future analysis of the Mg content in the PMgA.

Heating of magnesium hydroxide gives magnesium oxide with a 69% weight retention. Magnesium hydroxide was synthesized and TGA thermal analysis performed on it. A 70% weight retention, in agreement with the predicted value, was found.

$$Mg(OH)_2 \text{ ---> } MgO + H_2O$$

TGA weight retention results are consisted with that expected for a product containing a combination of structures 1 and 2, above.

One synthesis was repeated to access the reproducability. Further, for many of the synthesis two apparently different textured materials are produced. We also analyzed these two different textured materials for one sample. One texture is a white powder and the other is a transparent film. TGA thermograms, FT-IR and elemental analysis results are the same for all of these products. For instance, for the two textures of material and a replicate of this same reaction, the percentages of Mg are 10.52% (powder), 10.34 (film) and 10.34 (duplicate).

Structures for the chelation of Mg can be of two major forms referred to as bridged (such as 1 and 2) or non-bridges as shown in 3. These two structures can be identified by the location of IR bands associated with the carbonyl stretch. Thus, non-bridging structures show a band in the range of 1610 to 1650 corresponding to the asymmetric stretching and a band about 1360 corresponding to the symmetric stretch. By comparison, bridging structures exhibit a band about 1570 corresponding to the asymmetric stretching vibration and a band about 1420 corresponding to the symmetric stretch.

Structure 3

```
-HCH-CH-
      |
      C=O
      |
      O
      |
      Mg
      |
```

Strong bands are found at 1556 and 1421 consistent with the magnesium being present in a bridged structure.

Mass spectral analysis results are consistent with the proposed structures. As expected, there is a high similarity between the ion fragments found for PAA and those produced from PMgA. This is probably a consequence of two features. First, both structures are similar and will produce major ion fragments corresponding to the general structures C_xH_y and $C_xH_y(COOH)_z$ (and the corresponding deprotonated form). Second, salts are generally relatively thermally stable particularly under the conditions typically employed for mass spectroscopy. Because of the second factor, we have found that generation of metal-containing ion fragments generally requires rapid heating to above 400 C. Rapid heating to about 450 C was used to encourage the formation of metal-containing ion fragments. In assigning ion fragments as containing Mg, we compared the relative intensities of masses comparing the relative intensities of masses corresponding to reasonable Mg-containing ion fragments. Assignments of the most intense ion fragments are given in Table 1. These assignments contain a number of Mg-containing ion fragments. Ion fragments consistent with the formation of mono- and di-chelated Mg are present. Ion fragments at (all ion fragments are given in m/e=1; Daltons) 112 (COO-Mg-OOC) and 125 (COO-Mg-OOC-CH) are present consistent with the formation of di-chelated products.

Characterization of Poly(Zirconyl Acrylate)

Zirconyl chloride reacts directly with sodium hydroxide forming zirconium hydroxide.

$$ZrOCl_2 + NaOH \text{ ---> } ZrO(OH)_2 \text{ or } Zr(OH)_4 + H_2O$$

In truth, no zirconium hydroxide, $Zr(OH)_4$, exists (9). Instead, the precipitate is best described as $ZrO_2 \cdot H_2O$ or $ZrO(OH)_2 \cdot H_2O$.

Upon dissolving in water, zirconyl chloride forms ZrO^{+2} which is the zirconium-containing moiety added to the polymer. The solution is highly acidic with a 0.1 M solution of zirconyl chloride giving a pH of about 1.5. This is probably a result of a reversible reaction of ZrO^{+2} with water forming protons.

$$ZrOCl_2 \text{ ---> } ZrO^{+2} + 2\ Cl^-$$

$$ZrO^{+2} + H_2O \text{ ---> } ZrO(OH)^+ + H^+$$

Unlike Mg^{+2}, ZrO^{+2} reacts with PAA, partially neutralized PAA and fully neutralized PAA, ie. PNaA. The product between PAA and zirconyl chloride probably results because ZrO^{+2} is a stronger Lewis acid than the magnesium ion because of the presence of the electron-withdrawing oxygen on the zirconyl chloride giving it a "Lewis acid" strength sufficient to displace the carboxylate proton. The yield is low probably due to the formation of the initial product that will now have positively charged sites that discourage further reaction with positively charged chemical species.

```
-HCH-CH-    +  ZrO+2   --->   -HCH-CH-    + H-
     |                             |
   O=C-O-H                       O=C-O-ZrO-
```

Yield increases as the amount of sodium hydroxide, and associated extent of neutralization, increases reaching a maximum corresponding to an equal molar amount of reaction sites (that is, a molar amount of carboxylate groups to ZrO^{+2} groups of 2:1). This is a consequence of the increased ease with which the ZrO^{+2} is able to approach and react with a negatively charged chain and with the negatively charged (neutralized) carboxylate sites. Product yield greatly decreases as the amount of sodium hydroxide is in slight excess of the equivalent amount. But the weight of the product again increases as larger excesses of sodium hydroxide are added presumably because of the reaction of sodium hydroxide with the zirconyl ion to form zirconium hydroxide.

Table 1. Fragment Analysis of PMgA and PAA

m/e	poly(magnesium acrylate)		poly(acrylic acid)	
	%INT	Fragments	%INT	Fragments
55	73	C_4H_7	70	C_4H_7
56	25	C_4H_8	23	C_4H_8
57	83	MgO(OH) + C_4H_9	68	C_4H_9
60	17	H_3C-CO_2H	53	H_3C-CO_2H
67	35	C_5H_7	18	C_5H_7
68	31	CO_2Mg	18	C_5H_8
69	100	CO_2HMg	100	C_5H_9
70	33	CO_2HMg^{25}	22	C_5H_{10}
71	52	CO_2HMg^{26}	34	C_5H_{11}
73	18	$C_2H_4CO_2H$	58	$C_2H_4CO_2H$
81	94	$HC-CO_2Mg$	39	C_6H_9
82	29	H_2C-CO_2Mg	14	C_6H_{10}
83	46	H_3C-CO_2Mg	28	C_6H_{11}
85	30	$H_3C-CO_2Mg^{26}$	20	$C_3H_3CO_2H$
93	21	$H-CO_2Mg_2$	0	N/A
95	44	$C_2H_3CO_2Mg$	15	C_7H_{11}
96	18	C_7H_{12}	17	C_7H_{12}
97	37	$C_2H_5CO_2Mg$	18	C_7H_{13}
107	15	$C_3H_3CO_2Mg$	5	C_8H_{11}
109	24	$C_3H_5CO_2Mg$	7	C_8H_{13}
111	21	$C_3H_7CO_2Mg$	11	C_8H_{15}
121	20	$C_4H_5CO_2Mg$	6	C_9H_{13}
123	20	$C_4H_7CO_2Mg$	6	C_9H_{15}
136	16	$C_5H_8CO_2Mg$	7	$C_{10}H_{16}$
137	27	$C_5H_9CO_2Mg$	13	$C_{10}H_{17}$
149	28	$C_{11}H_{17}$	27	$C_{11}H_{17}$

These reactions are probably a result of a group of competing equilibrium reactions that favors the neutralization of the carboxylic groups through reaction with sodium hydroxide with PAA instead of reaction between ZrO^{+2} and sodium hydroxide. Subsequent reaction between ZrO^{+2} and PNaA occurs during this time. In excess, sodium hydroxide reacts with ZrO^{+2} forming zirconium hydroxide.

Unless noted otherwise, the following discusses only the product formed from a 2:1 molar ratio of PNaA and ZrO^{+2}.

Infrared spectroscopy is consistent with the product being largely of the bridged structure with strong bands at 1541 and 1454. Small bands are found at 1699 and 1344 consistent with the presence of some un-reacted acid sites.

Elemental analysis shows 44% zirconium. Calculated values for possible structures are $PZrOA_2$ (37%), PZrOClA (43%) and PZrO(OH)A (47%). The found value is consistent with the product being a combination of these structures.

Thermal analysis is also consistent with the product being a combination of structures as noted above. Thus, for the product formed from a 1:2:2 ratio of PAA:NaOH:$ZrOCl_2$ a 36% residue (ZrO_2) is found. A calculated value for the structure $PZrOA_2$ is 49 % and for PZrO(OH)A it is 63%.

Finally, mass spectroscopy was not of assistance in characterizing the structure. Presumably, the ZrO moiety is strongly enough bound so that ion fragments containing this moiety are not found.

Preparation of Mg-PSZ Ceramics

Pre-ceramic material was made using material produced through reaction of 0.8 M (10-Mole % Magnesium sulfate and 90-Mole-% $ZrOCl_2$) metal ions with 0.4 M PAA that had been neutralized by addition of 0.4 M of NaOH. The material was processed in the usual fashion using heating to 1700 C along with associated grinding, etc.

For good Mg-PSZ ceramics, a diameter shrinkage of 18% is found. For the above described product the diameter shrinkage was only 15%. The relative percentage density for the PAA-derived material was only 80.3% of that of Mg-PSZ itself.

Scanning Electron Microscopy, SEM, showed mini-cracks but with a grain size of only 10 um compared with a well sintered Mg-PSZ ceramic showing a larger grain size of about 50 um.

Energy Dispersive Spectroscopy, EDS, was performed on the product. From EDS areal scanning the magnesium content is 1.4% while the zirconium content is 98.6% by atom number fairly far away from the desired ratio of about 10:90.

Even with this low amount of magnesium, the pre-ceramic PAA-derived material allowed the formation of Mg-PSZ ceramics with values approaching those of good Mg-PSZ ceramic materials.

REFERENCES

1. Ceramics (Japanese), 16, 570 (1981).
2. A. C. Pierre, Ceramic Bulletin, 70(8), 1281 (1991).
3. M. K. Agarwala, J. Amer. Ceram. Soc., 75(7), 1975 (1992).
4. D. L. Bourell, J. Amer. Ceram. Soc., 76(3), 705 (1993).
5. D. C. Bradley, R. C. Mehrotra and D. P. Gaur, "Metal Alkoxides", Academic Press, NY, p.338, 1978.
6. K. S. Mazdiyasni, U. S. A. F. Tech. Doc. Rept., ASD-TDR-63-332, May (1963).

7. I. M. Thomas, U. S. Pat. 3,799,909 March 26 (1974).
8. G. Zhang, Tianjin University, Masters Thesis (1990).
9. F. A. Cotton and G. Wilkinson, "Advanced Inorganic Chemistry' Interscience, NY, pp 766-768, 1962.

PREPARATION OF FUNCTIONALIZED POLYMERS BY REACTIONS OF POLY(VINYLBENZYL CHLORIDE)

Kristin L. Thunhorst, Richard D. Noble, and Christopher N. Bowman*

Chemical Engineering Department
Engineering Center, ECCH 111
University of Colorado
Boulder, Colorado 80309-0424

ABSTRACT

Functionalized polymers are of interest in a variety of applications including but not limited to fire retardants, selective sorption resins, chromatography media, controlled release devices and phase transfer catalysts. This research has been conducted in an effort to functionalize a polymer with a variety of different reactive sites for use in membrane applications. These membranes are to be used for the specific separation and removal of metal ions of interest. A porous support was used to obtain membranes of a specified thickness with the desired mechanical stability. The monomer employed in this study was vinylbenzyl chloride, and it was lightly crosslinked with divinylbenzene in a photopolymerization. Specific ligands incorporated into the membrane film include dimethyl phosphonate esters, isopropyl phosphonate esters, phosphonic acid, and triethyl ammonium chloride groups. Most of the functionalization reactions were conducted with the solid membrane and liquid reactants, however, the vinylbenzyl chloride monomer was transformed to vinylbenzyl triethyl ammonium chloride prior to polymerization in some cases. The reaction conditions and analysis tools for uniformly derivatizing the crosslinked vinylbenzyl chloride / divinyl benzene films are presented in detail.

INTRODUCTION

Poly(vinylbenzyl chloride) (VBC) is an ideal starting material onto which a variety of functional groups can be attached through relatively simple reactions and mild reaction conditions. Functionalized polymers are of interest in a variety of applications including but not limited to fire retardants, selective sorption resins, chromatography media, controlled release devices and phase transfer catalysts. An example of the wide applicability of functionalized polymers is provided by trimethyl ammonium functionalized poly(VBC).

Based on the identity of the anion associated with the charged polymer, it can catalyze reactions ranging from oxidation of diol compounds to conversion of alkyl halides to aldehydes[1]. Alexandratos *et al.* have synthesized many functionalized polymers for a variety of ion exchange, complexation, and reduction reactions, some of which can be simultaneous[2]. Several other researchers have also synthesized phosphorous-functionalized polymers for ion exchange and metal ion chelation[3,4].

The functionalized polymers prepared in this work were synthesized for use as membranes containing specific sorption ligands and/or membrane polymers with fixed charges on the backbone. Several phosphorous-containing polymers and one nitrogen-containing polymer were prepared. Based on accounts in the literature regarding sorption resins, the complexation of metal ions with a variety of functional groups was known to be a function of the pH. It was desirable to synthesize membranes with the same functional groups (tested as sorption resins) in hopes of obtaining a facilitated transport membrane. The goal of facilitated transport is to manipulate environmental conditions on each side of a membrane (e.g. pH) to drive solute from a low concentration (feed side) to a greater concentration (receiving side). The facilitated transport membranes would enjoy applications in catalyst recycling, wastewater treatment, drinking water treatment, even in metal ion analyses and detection. The specific identities of the ligands attached to the poly(vinylbenzyl chloride) (PVBC) backbone include methyl phosphonate ester (MPE), isopropyl phosphonate ester (IPE), phosphonic acid (PA) and triethyl ammonium chloride groups. This paper focuses on the syntheses and some of the characterization of the functionalized polymer films.

SYNTHESES

Membrane Preparation

The vinylbenzyl chloride (VBC) monomer used in this work was purchased from Aldrich (Milwaukee, WI) as an inhibited monomer mixture of the 3- and 4- isomers. The inhibitor was removed from the monomer by contact with a column of De-hibit 100 Ion Exchange Resin from Polysciences (Warrington, PA). The divinylbenzene used as a crosslinking agent was also purchased from Aldrich as an inhibited 55% solution of the isomers with 3- and 4-ethylvinylbenzene. Due to the extremely small amount of divinylbenzene used in the monomer mixtures, the inhibitor was not removed prior to polymerization. The remaining reactants and solvents include triethyl amine (99%), trimethyl phosphite (99+%), triisopropyl phosphite (95%), acetone, chloroform, and hexyl alcohol which were also purchased from Aldrich. The photoinitiator used to initiate the polymerizations was Irgacure 907, donated by Ciba Geigy (Hawthorne, NY).

Two membrane functionalization tactics were employed in this research. In the first case, solid membranes of crosslinked PVBC were synthesized and subsequently functionalized via reactions with liquid phase reactants. In the second case, the monomer was functionalized with the desired reactive groups and purified before co-polymerization with VBC and divinylbenzene. The only membranes prepared by the second method contained vinylbenzyl triethyl ammonium chloride (VBTAC). This method was possible because of the mild reaction conditions and the fact that the VBTAC was a solid after reaction. For all experiments conducted in this study, the amine monomer was co-polymerized with VBC at 37 mole percent. Attempts to functionalize VBC with the methyl phosphonate ester group prior to polymerization were not successful.

To form the membranes, the desired monomer components were combined and hexyl alcohol was added to the mixture until it comprised 20-50 weight percent of the total. The hexyl alcohol was added to monomer mixtures formed with VBTAC to solvate the liquid monomers (VBC and divinylbenzene) with the VBTAC monomer. Hexyl alcohol was also added to monomer mixtures which did not contain VBTAC, not for solvation, but for consistency in preparation schemes. The photoinitiator was added to the hexyl alcohol / monomer solutions so that it comprised approximately 2 weight percent. Divinylbenzene was included in the monomer mixture in amounts from 1.2 to 1.4 mole percent of monomers so that the polymers would be lightly crosslinked to prevent solubilization of functionalized polymers during transport experiments.

To achieve the mechanical strength needed for use as a membrane, a microporous polyethylene support was used. The support, obtained from 3M (Minneapolis, MN), was approximately 20 microns in thickness and was 83% porous. The liquid monomer mixtures were applied to the support with a wiped film method and subsequently polymerized with UV irradiation at 365 nm and 2.0 mW/cm^2 for 2 hours. After polymerization, a solid transparent film was obtained. To allow unreacted monomer and oligomer to leave the film, it was placed into chloroform for three days, with fresh chloroform exchanges once per day. After the chloroform exchanges, the film was allowed to dry in the chemical hood.

Composition of Membrane

To measure the amount of polymer incorporated in the support, weight gain measurements were initially used. Due to the small sample mass and the static associated with the thin films, this method was not extremely accurate and was abandoned later in the studies in favor of using a Perkin Elmer Differential Scanning Calorimeter (PE DSC7). We were able to use the DSC for composition measurements because the polyethylene support had a melting transition at 120°C with a characteristic heat of 29 calories/gram. The polymers incorporated into the pores of the polyethylene were crosslinked and exhibited no transitions in this region, thus the sample mass percent of polyethylene could be determined with the DSC. This determination was completed on all samples immediately after drying from the chloroform rinses, before any further reactions were performed on the film. The samples after polymerization were typically 20 to 30 microns thick (from Scanning Electron Microscopy investigation) and the pores of the support appeared to be completely filled. Based on DSC analyses, the membranes typically had 20 to 30 percent by mass polyethylene which corresponds well with the reported porosity.

Membrane Functionalization

Membranes which were prepared from crosslinked PVBC were typically functionalized by immersing the solid PVBC and polyethylene film into a liquid bath containing reactants. This procedure was followed to obtain membranes functionalized with the MPE, IPE and PA

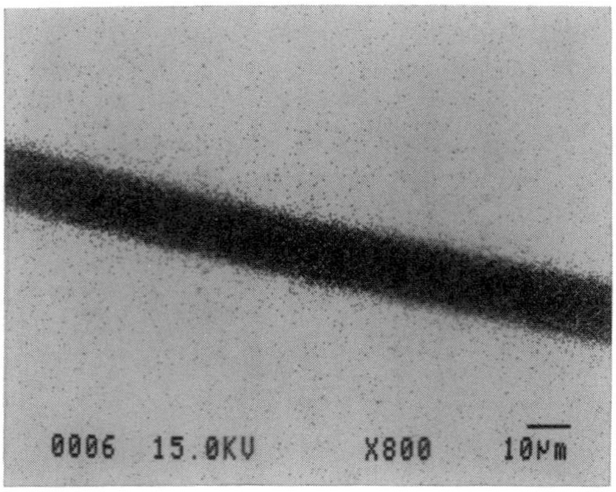

Figure 1. Cross-section of a membrane functionalized with MPE functional groups imaged by x-ray dot mapping of phosphorous. The sparse black dots represent the background level of the epoxy used to mount the sample, while the concentrated band represents the membrane cross-section.

groups. Because these films were to be used in membrane applications, it was desirable to achieve a uniform concentration of reactive sites across the cross-section of the functionalized film. The method employed to achieve this goal was to ramp the temperature of the reactant bath very slowly up to the temperature at which reaction occurred. This strategy allowed the liquid reactants to diffuse into the solid matrix prior to achieving the correct conditions for reaction. The uniformity of reactive sites across the membrane cross-section was verified by performing wavelength dispersive x-ray dot mapping of phosphorous on a JEOL Superprobe JXA-8600 on representative samples. One of the cross-sectional phosphorus scans has been included in Figure 1. The homogeneous concentration of phosphorous in the cross-section indicated that the diffusion resistance of the membrane did not cause displacement reactions at the membrane surface alone. This is unique with respect to other functionalization reactions on items such as cross-linked beads. Many bead derivatization reactions occur solely on the exterior surface of the cross-linked resin.

The reaction to append MPE functional groups was conducted in a three-necked round bottom flask with a water condenser mounted in the vertical neck. One of the necks of the flask allowed the use of a thermometer to constantly measure reactant bath temperature, while the second permitted the introduction of a nitrogen atmosphere. Nitrogen was not constantly flushed through the apparatus, it was introduced periodically. The top of the water condenser was closed with a ground glass stopper except at the times when the reaction vessel was being purged with nitrogen. The reason for the nitrogen atmosphere and the stopper on the top of the condenser was to avoid allowing atmospheric water into the reactor. The round bottom flask was heated with a heating mantle. Cold water was constantly allowed to flow through the condenser to prevent the volatilization and loss of the trimethyl phosphite reactant.

More than 100 times excess trimethyl phosphite was charged to the reactor with the PVBC film to ensure that an adequate amount of reactant was present despite side reactions. The reaction vessel was purged for 3 minutes with nitrogen after which the nitrogen purge and the condenser exit were closed. The temperature of the reaction bath was ramped over

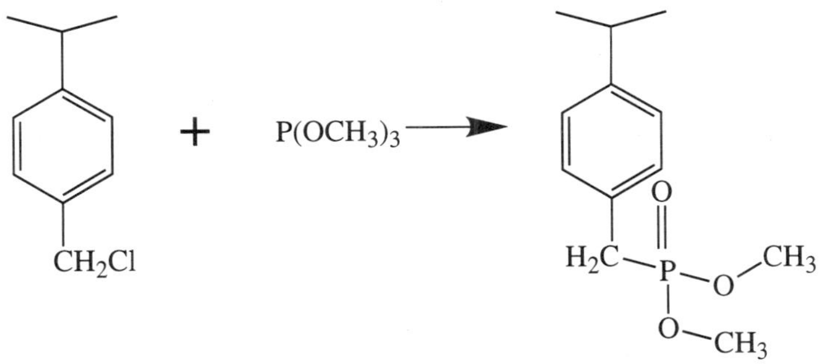

Figure 2. Representation of the reaction to form the MPE functionalized membranes.

three hours from room temperature and stabilized at approximately 100°C. The reaction was allowed to proceed from 1 to 5 days and samples were periodically withdrawn throughout this period. Whenever the reaction vessel was opened to remove a sample, the nitrogen purge was repeated. The reaction (shown in Figure 2) is similar in mechanism to an S_N2 reaction and is referred to as Michaelis-Arbuzov chemistry[5]. Figure 2 contains a simplified representation of the polymeric sample, since in reality the product was a crosslinked network contained in a polyethylene support. When the reaction was complete, the films were soaked in acetone or chloroform for three days with fresh solution exchange each day to remove unreacted phosphite species.

The percent of the benzyl chlorines replaced in the reaction was referred to as percent functionalization and was controlled via the reaction time. Figure 3 contains the data of percent functionalization versus time for a number of functionalized membranes.

In general, longer reaction times led to greater values of functionalization. The percent functionalization was measured by weight gain of the films as determined on a balance. This method of measurement was inherently inaccurate due to the physical nature of the reaction. Elemental analyses were performed on a number of membranes to assess the error in the weight gain measurements of percent functionalization. Based on these elemental analyses,

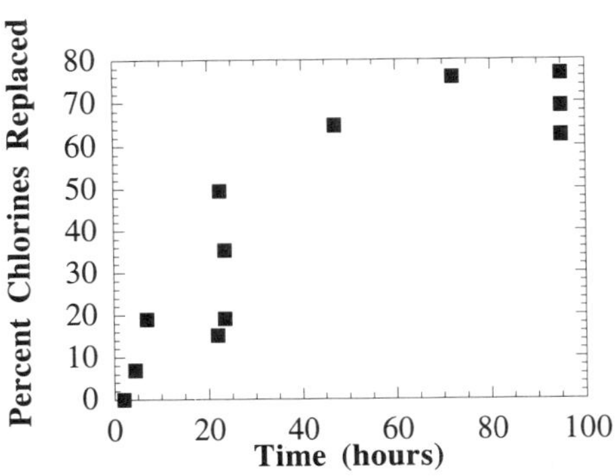

Figure 3. The percent of benzyl chlorines replaced in a reaction with trimethyl phosphite as a function of reaction time (measured by weight gain). The reaction time is measured from the time the reaction bath reached 100°C.

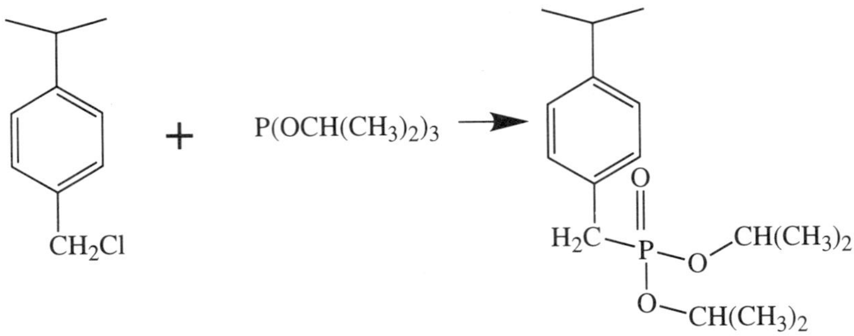

Figure 4. Simplified representation of reaction used to form IPE functionalized membranes.

the error in percent functionalization values determined by weight gain on the balance was approximately 14%. The values reported for percent functionalization are expected in all cases to be less than the actual values of percent functionalization. This is due to the physical nature of the reaction, during which small pieces of the membrane could have been removed by tearing. The loss of mass would cause the results to be skewed toward lower values of functionalization.

The IPE functionalized membranes were generated using the same synthesis apparatus as the trimethyl phosphite reaction, but with slightly different reaction conditions. Anhydrous ethanol was used as a solvent for this reaction, and an excess of triisopropyl phosphite was employed. The reaction temperature for this reaction was approximately 78°C as opposed to the 100°C temperature used in the reaction with trimethyl phosphite, but temperature ramping was used in both instances. A large excess of the triisopropyl phosphite coupled with the several-day reaction time provided complete conversion of the benzyl chlorines on the polymer film to the IPE groups. Once the reaction was complete, the unreacted phosphite species were removed by the method described previously. The reaction mechanism to synthesize the IPE functionalized polymer was essentially the same as that to synthesize the methyl species. The polymer representations are simplified in Figure 4, since they actually represent the lightly crosslinked polymers within a polyethylene support.

The IPE functionalized membranes were generated as an intermediate step to obtaining the PA functionalized films. Our attempts to hydrolyze the MPE to the PA group were not successful, however the IPE was successfully converted to the acid form. To synthesize the acid form, the IPE functionalized membranes were refluxed in concentrated (37 wt%) hydrochloric acid for 17 hours.

Monomer Functionalization

The only functionalization reaction which was not performed on the crosslinked PVBC film was the one which resulted in VBTAC. VBTAC-containing membranes were synthesized by a co-polymerization of VBC, divinylbenzene and VBTAC. The VBC monomer was functionalized with the quaternary amine groups prior to polymerization. This was possible because the reaction and purification conditions to produce the VBTAC monomer were quite mild and did not induce polymerization of the VBC.

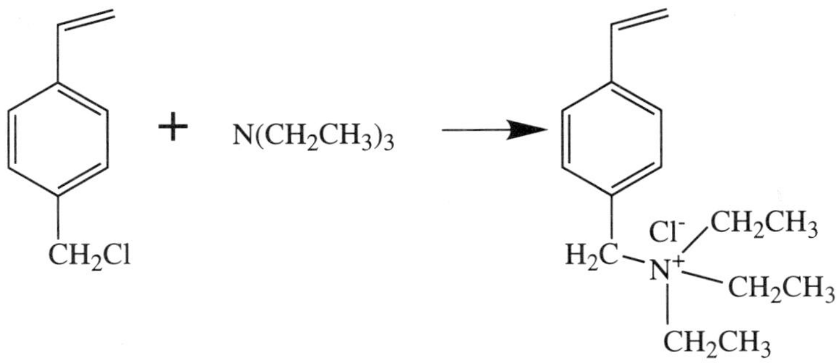

Figure 5. Simplified representation of the reaction of VBC to the VBTAC monomer.

To functionalize VBC with the quaternary amine groups, triethyl amine (TEA) was used. The TEA was added (at 100% molar excess) to a round bottom reaction flask with VBC from which the inhibitor had not been removed. The apparatus used for this synthesis was very similar to that used to synthesize the MPE functionalized films, except a water bath replaced the heating mantle and a magnetic stir bar was used to agitate the reactor contents throughout the reaction. The water bath temperature was ramped slowly to bring the reactants to 40°C and to avoid temperature overshoot which may have resulted in polymerization of the monomer. After an hour at 40°C, a white solid precipitate (VBTAC) started to form at the bottom of the reactor. As the reaction time increased, so did the amount of product in the reactor bottom, until the stir bar was unable to further agitate the wet solid. After 71 hours of reaction, 72% of the VBC had been converted to the desired product (measured through the use of NMR described in Characterization section).

The VBTAC monomer was insoluble in acetone despite the solubility of both reactants. The product was thus cleaned and purified with several washes in acetone, after which it was filtered and dried. The product purity was greater than 98% after the acetone washes based on NMR analyses. The functionalized monomer was stored in a dark chemical refrigerator and the same batch was used throughout the studies. The VBTAC monomer was co-polymerized with VBC as described in the membrane preparation section of this article. The reaction is represented in Figure 5.

Characterization

Fourier Transform Infrared Spectroscopy (FTIR) provided a convenient tool by which reaction success was qualitatively evaluated. The spectrum of unfunctionalized PVBC on the polyethylene support is included in Figure 6a. Because of the intense absorbances of the PA, MPE, and IPE species, FTIR proved to be a convenient tool to investigate the progression of the syntheses. Note the changes in the spectra as the membranes were functionalized especially in the range of 900-1250 wavenumber for the MPE and IPE functional groups[6,7]. The PA functionalized membranes have several other very broad characteristic absorbances[7]. Figures 6b, 6c, and 6d contain the FTIR spectra of the membranes functionalized with the phosphorous species. Note that the intense absorbances around 3000 wavenumber are derived from the polyethylene support.

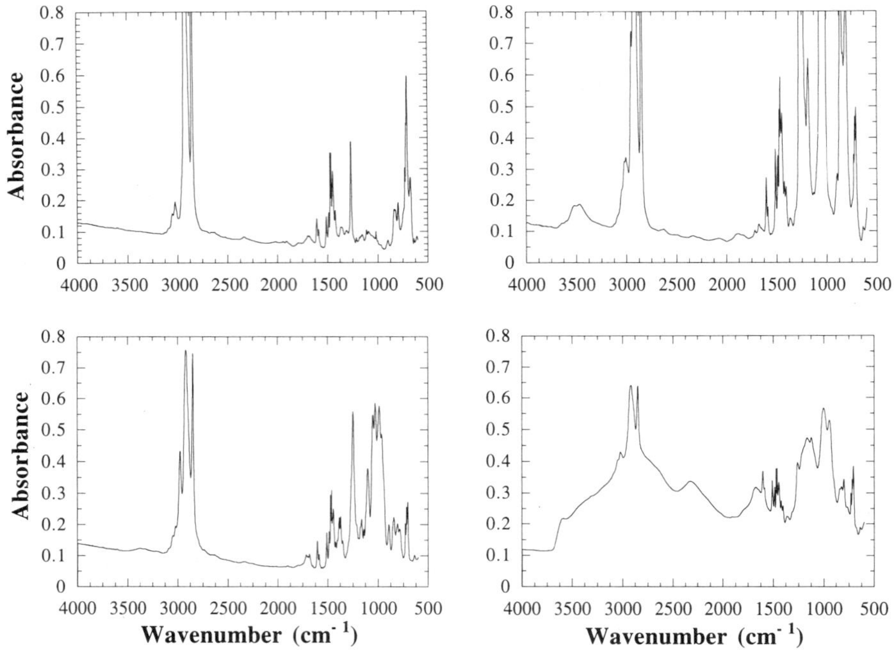

Figure 6. Plots of Absorbance vs. Wavenumber for a variety of membranes synthesized in this work, each includes the polyethylene support: a) PVBC membrane, b) membrane functionalized with the MPE groups, c) membrane functionalized with the IPE groups, and d) membrane functionalized with PA groups.

Quantitative measurement of the percent functionalization was assessed with the DSC, which replaced the initial weight gain measurements completed with a balance. The DSC provided weight gain information indirectly via the heat of melting of the polyethylene support. As mentioned earlier, after polymerization the membranes were heated in the DSC to obtain the mass percent polyethylene. The heat of melting was determined for the same membrane samples after functionalization. The weight fraction of polyethylene was altered in the samples during the reaction since the functional groups being added to the polymer had significantly greater mass than the chlorine which was present initially. By comparing the sample weight percent polyethylene before and after reaction, a system of equations was developed which yielded the percent of chlorine sites that were replaced by the phosphorous-containing ligand. Given the molecular weight of both the chlorine leaving group and the reactive group being added to the polymer backbone, in addition to several linear equations involving species not affected by reaction, the percent functionalization was calculated.

The technique of using the DSC to determine percent functionalization was subject to error, however, with polymer films that had a propensity to retain water. In this study, the membranes functionalized with the PA group were hydrophilic enough that they contained water after the acidic hydrolysis reaction. Despite the DSC operating temperatures which exceeded 130°C, water was still retained in the membranes. The presence of such strongly sorbed water in the films resulted in an artificially low measured mass percent of polyethylene in the samples, and thus calculated functionalization values exceeding 100%. Thus, this technique must be carefully applied to samples which strongly retain or sorb

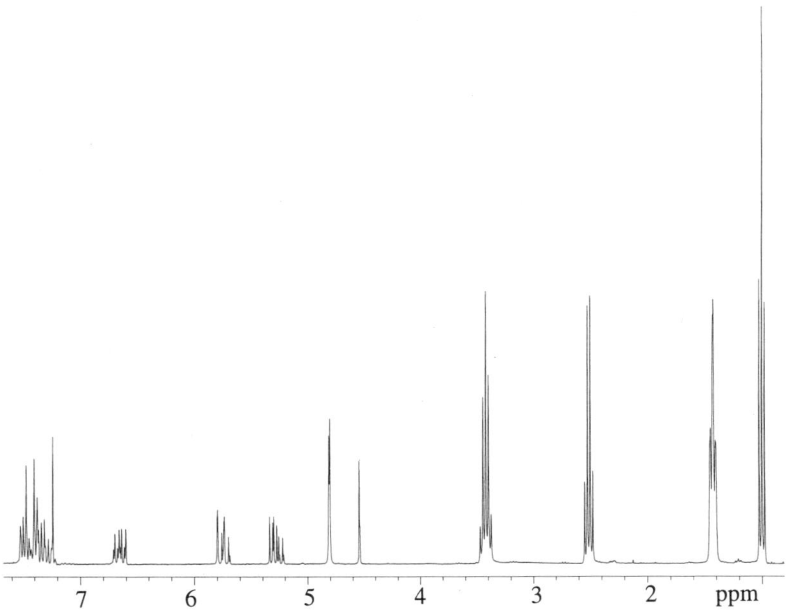

Figure 7. Peak positions in the proton NMR spectrum of the reaction mixture with VBC, TEA, and VBTAC.

compounds that are not a part of the actual polymer sample of interest. The MPE and IPE functionalized samples were not subject to this kind of error in measurement, and the percent functionalization values calculated for them were credible.

The functionalization reaction of the VBC monomer to VBTAC was followed with proton NMR. The spectrometer used was a Varian VXR-300S, with deuterated solvents including deuterated chloroform, and deuterium oxide. The reaction conversion was monitored by samples which were periodically withdrawn and run on the NMR. The reaction mixture contained both reactants, TEA and VBC, as well as the product. The ethyl amine hydrogens of the unreacted TEA appeared at a lower chemical shift than the ethyl amine hydrogens of the VBTAC. The benzyl hydrogens of VBC were at a lower chemical shift than the benzyl hydrogens of the VBTAC. The conversion of VBC to VBTAC was measured by integrating and comparing the two benzyl hydrogen peaks. By the same method, the purity of the solid product after acetone washes was assessed. A sample of the spectrum is included in Figure 7 and is referenced to $CDCl_3$ at 7.24 ppm.

TRANSPORT RESULTS

Experiment Description

The functionalized membranes were tested in ion transport experiments to evaluate the effect of type and level of functionalization[8]. The membranes were contacted on one side with a feed solution containing 1 mM concentrations of sodium, zinc, ferric and/or neodymium nitrate salts. The receiving solution in contact with the other side of the membrane typically contained 1 M nitric acid. The transport of the metal ions through the membrane was measured via samples withdrawn at specific time intervals and analyzed with Inductively Coupled Plasma Atomic Emission Spectroscopy (ICP-AES). The MPE, IPE, PA, and VBTAC functionalized membranes were each tested in ion transport experiments. Membranes with varying functionalization of the MPE, IPE and PA groups were evaluated. Experiments were also run with PVBC unfunctionalized membranes. The membrane area available for ion transport was 2.8 cm^2.

Transport Results

Flux values were calculated by linearizing the long term receiving side metal ion concentration profiles versus time. The measured flux values for sodium, zinc and ferric ions through the unfunctionalized PVBC membranes were measured as $3*10^{-11}$ mol/cm^2min in one experiment and between 0 and $7*10^{-12}$ mol/cm^2min when the experiment was repeated. These data are considered to be acceptable replicates because of the extremely low flux values through the unfunctionalized membranes. Neither the IPE nor the PA membranes exhibited greater ion flux than the unfunctionalized PVBC membranes, despite experiments run with membranes of high functionalization. The MPE functionalized membranes, however, did exhibit increases in the ion flux as the functionalization was increased. The ion flux increase was highly nonlinear with functionalization, displaying behavior similar to what has been termed a *percolation threshold* by other researchers[9].

To assess the impact of the functionalization on the ion transport, the normalized flux values for each of the ions were calculated[8]. The normalized flux is defined as the flux of the ion through a functionalized membrane divided by the flux of the same ion through an unfunctionalized membrane. For MPE functionalized membranes and the ferric ion, the normalized flux values were approximately 1 between 1 and 30 percent functionalization and rose in a sudden manner to 10-15 at 65-69 percent functionalization. The effects were more pronounced when the flux of the sodium ion through MPE functionalized membranes was considered. Normalized flux values of the sodium ion were approximately 1, at less than 30 percent functionalization, and ascended dramatically to 35-80 at 65-69 percent functionalization. The authors suspect that the functionalization of the membranes altered the swelling and thus, the glass transition temperature of the swollen, lightly crosslinked films, resulting in dramatic flux increases[8].

CONCLUSIONS

PVBC has proven to be a versatile starting material which is polymerized and functionalized with a variety of simple chemical reactions. Several phosphorous-containing species were attached at the position of the benzyl chlorine in the lightly crosslinked film of PVBC. These species include the methyl phosphonate ester, isopropyl phosphonate ester and phosphonic acid. The reactions with phosphorous-containing species took place on a solid film of the PVBC with liquid reactants. Homogenous replacement of the benzyl chlorines with the phosphorous-containing ligands across the membrane cross-section was verified. An addition reaction with triethyl amine on the VBC monomer provided a nitrogen-containing functionalized monomer, vinylbenzyl triethyl ammonium chloride (VBTAC). The VBTAC was successfully co-polymerized with VBC and the resulting membranes were tested in ion transport experiments. Ion transport experiments were also performed with membranes containing each of the MPE, IPE, and PA functional groups in varying amounts. The flux of sodium, zinc and ferric nitrate salts through the membranes was found to be affected to a large extent by the level of functionalization with the MPE ligand.

ACKNOWLEDGMENTS

The authors would like to thank NSF for its support of this work through a graduate Student Fellowship to KLT and both the I/U CRC Center for Separations Using Thin Films and the Colorado Advanced Technology Institute for their financial support of the project. The authors would also like to acknowledge 3M for donation of the microporous polyethylene support used in this work.

REFERENCES

1. A. Akelah, and A. Moet. *Functionalized Polymers and Their Applications*; Chapman and Hall, New York (1990).
2. S.D. Alexandratos, D.W. Crick, D.R. Quillen and C.E. Grady, Chapter 14. Polymeric ligands for ionic and molecular separations, in: *Environmental Remediation Removing Organic and Metal Ion Pollutants,* G.F. Vandegrift, D.T. Reed, and I.R. Tasker, eds., American Chemical Society, Washington D.C. (1992).
3. G.C. Daul, J.D. Reid, and R.M. Reinhardt, Cation exchange materials from cotton and polyvinyl phosphate, *Ind. Eng. Chem.,* 46:1042 (1954).
4. H. Egawa, T. Nonaka, H. Maeda, Studies on selective adsorption resins. XXI. Preparation and properties of macroreticular chelating ion exchange resins containing phosphoric acid groups *J. App. Polym. Sci.,* 30:3239 (1985).
5. R. Engel. *Synthesis of Carbon-Phosphorous Bonds*; CRC Press Inc., Boca Raton (1988).
6. M.E. Grayson and E.J. Griffith, eds., *Topics in Phosphorous Chemistry*; Interscience Pub., New York (1969).
7. L.C. Thomas. *Interpretation of the Infrared Spectra of Organophosphorous Compounds*; Heyden and Son Ltd., New York (1974).
8. K.L. Thunhorst, R.D. Noble, and C. N. Bowman, Transport of ionic species through functionalized poly(vinylbenzyl chloride) membranes, *J. of Membrane Sci.,* 128:183 (1997).
9. E.L. Cussler, R. Aris, and A. Bhown, On the limits of facilitated diffusion, *J. Membrane Sci.,* 43:149 (1989).

GRAFT COPOLYMERIZATION OF VINYL MONOMERS ONTO POLYMERS CONTAINING STYRENE

David Jiang and Charles A. Wilkie

Department of Chemistry
Marquette University
P. O. Box 1881
Milwaukee, WI 53201

INTRODUCTION

The grafting of monomer onto polymers has been initiated by photochemical, high-energy radiation, and chemical processes. As long as a radical or ionic site may be generated on the polymer and the monomer is sufficiently active, graft copolymerization may occur. Two photochemical processes have been described. The first involves the anthracene sensitized formation of singlet oxygen and its addition to an allylic position in a butadiene containing polymer to give a hydroperoxide; this hydroperoxide may be thermally decomposed so that monomer addition may occur at that site.[1,2] This is shown as Scheme 1. In the second procedure[3,4] the irradiation of benzophenone produces a radical triplet state which can abstract a hydrogen atom; this is shown as Scheme 2. In the first case, it has been assumed that addition of singlet oxygen occurs at the weakest carbon-hydrogen bond, the allylic C-H, and in the second case the assumption is that the radical adds to the double bond. In neither case has the site of grafting been definitively proven.

High-energy radiation is a non-selective process in that the energetic radiation will remove atoms and generate radical sites more or less at random. It is not possible to predict the site at which graft copolymerization may occur.

Chemical processes are far more varied and may involve either the formation of radicals or ions along a polymeric backbone. Both cationic processes[5] as well as radical processes have been widely used for graft copolymerization of vinyl monomers onto various polymers. Radical graft copolymerization has been reported for many polymers including styrene-butadiene block copolymers,[1,2,6] and acrylonitrile-butadiene-styrene terpolymer, ABS.[3,7-9]

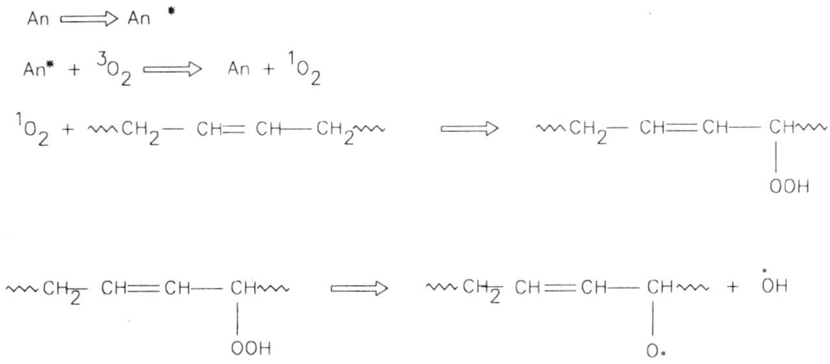

Scheme 1. Anthracene photosensitized production of radicals on butadiene fragments

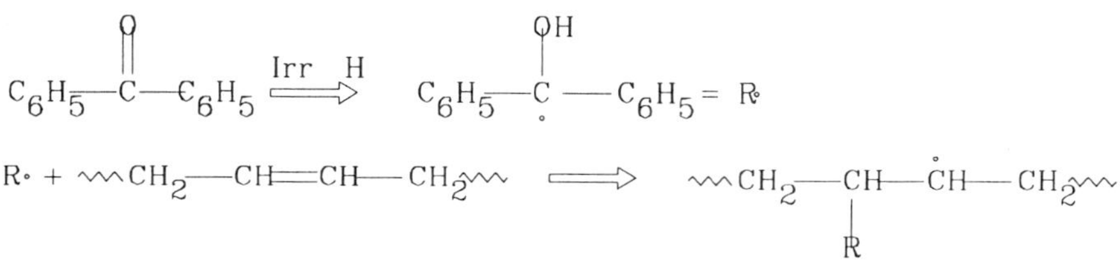

Scheme 2. Benzophenone sensitized of radicals along a butadiene chain

The usual proof of graft copolymerization is based upon changes in solubility between the starting polymer and the graft copolymer. Some workers have measured some physical property, such as contact angle or wetability, and used a change in that parameter as an indication of the formation of a graft copolymer.[10-15] In previous work from this laboratory on the interaction between poly(ethylene terephthalate) and vinyl monomers such as methacrylic acid, methyl methacrylate, and styrene mediated by various initiators, including BPO, AIBN, and some transition metal systems, it has been shown that graft copolymers are never formed but instead the vinyl monomer is hompolymerized and becomes entangled within the polymeric framework.[16,17] Since it has become physically, but not chemically, incorporated within the polymer, the physical properties, such as contact angle or wetability, are modified even though there is no chemical attachment. Spectroscopic evidence of a change in the polymeric structure is necessary to truly prove that a graft copolymer has been formed.

In this paper graft copolymerization onto both polystyrene[18] and styrene-butadiene block copolymer[19,20] will be discussed. It will be shown that radical processes do not permit the addition of monomers onto polystyrene and that one must use anionic initiation in order to cause the graft copolymerization reaction to occur. This of necessity means that the graft reaction occurs only on the butadiene portion of the styrene-butadiene block copolymer when the graft copolymerization reaction is initiated by radicals. The site on the polymer to which the vinyl monomer attaches has been spectroscopically identified.

EXPERIMENTAL

Materials

The vinyl monomers used in this study, methacrylic acid, methyl methacrylate, acrylic acid, methyl acrylate, and acrylonitrile, as well as the solvents, initiators, and polystyrene were supplied by the Aldrich Chemical Company. The styrene-butadiene block copolymer was supplied by Shell as Kraton D1102, known as SBS; this contains approximately 75% butadiene and 25% styrene.

Radical Graft Copolymerization onto Polystyrene

The graft copolymerization of acrylonitrile onto polystyrene was attempted using benzoyl peroxide, di-*t*-butylperoxide, and 2,5-dimethyl-2,5-di-(*t*-butylperoxy)hexane as initiators. In all cases no increase in mass of the polystyrene was observed. Attempts were also made to test whether the polystyryl radical was ever formed by combining the initiator and the polymer or the initiator, polymer and a nitroxide radical trap. In the first case the formation of a radical must lead to cross-linking of the polymer and in the second case the polystyryl radical will be trapped by the nitroxide.[18]

Photochemical Grafting Scheme

Anthracene was introduced into a film by sorption from a methanol solution; the concentration of anthracene in the film could be controlled by variation of the concentration in the solution and the concentration could be determined by ultraviolet spectroscopy. After irradiation hydroperoxides are formed which are thermally decomposed in the presence of a vinyl monomer leading to the formation of a graft copolymer. The films were recovered and washed thoroughly with water to remove any homopolymer. The amount of graft copolymer which was produced was determined by a comparison of the mass of starting material with the mass of the final film.[2,7]

Chemical Initiation of Graft Copolymerization onto Styrene-Butadiene Block Copolymer

Graft copolymerization of monomers onto SBS[19,20] was performed by dissolving a sample of the polymer in a solvent and adding an initiator and monomer to this solution. The temperature of the reaction, reaction time, and concentrations of monomer, initiator, and polymer were all varied. The reactions leads to both the formation of the homopolymer of the monomer as well as graft copolymer; in some cases unreacted starting material is also recovered. These three components may be separated by solubility differences between the homopolymers and the graft copolymer. The solvents which were used depend upon the particular monomer which has been graft copolymerized. When methyl methacrylate is used, the homopolymer of methyl methacrylate, the backbone SBS, and the graft copolymer are all soluble in the reaction solvent, chloroform, but only poly(methyl methacrylate) is soluble in acetone; this makes it possible to separate poly(methyl methacrylate) from the reaction mixture. When the monomer is methyl acrylate, the graft copolymer will dissolve to some extent in acetone but it is not soluble in a 10% solution of water in acetone but poly(methyl acrylate) will dissolve in this solvent. For both acrylic acid and methacrylic acid, the homopolymer of the monomer is soluble in both methanol and water and the backbone polymer, SBS, is chloroform soluble while graft copolymer will not dissolve in either methanol, water nor chloroform. In order to be sure that all homopolymer has been separated from the graft copolymer, the recovered mass was typically treated five times, alternating between a solvent for SBS and a solvent for the homopolymer. Infrared spectroscopy was used to determine the efficacy of the separation.

Identification of the Site of Graft Copolymerization

The insoluble material is assumed to be the graft copolymer and this is verified by infrared spectroscopy. For grafting onto the butadiene portion of a copolymer, the C-H out-of-plane bending vibrations as well as the olefin C-H stretching vibration are most useful. The graft copolymer of acrylonitrile onto polystyrene cannot be analyzed by infrared spectroscopy since the only change would be in the C-H overtone region and these bands are too weak to permit interpretation.

RESULTS AND DISCUSSION

Graft Copolymerization onto Polystyrene

Attempts at Radical Graft Copolymerization onto Polystyrene. An examination of the polystyrene structure reveals the presence of a benzylic hydrogen atom which should be easily lost to produce a tertiary radical and one would think that this reaction may proceed easily. Several attempts have been made with a variety of initiators to remove this hydrogen and thus effect the addition of monomers at this site but none of these attempts have been successful. In order to better understand this reaction, attempts were made to form and trap the radical with a nitroxide radical trap but these were likewise unsuccessful even though one may trap ethylbenzene under the identical conditions. The non-observation of the trapped radical of polystyrene may be attributed to a) too low a concentration of nitroxide - trapped polymer; b) the inability of the nitroxide to interact with the polymeric radical; c) an unusual stability of the polystyryl radical; or d) an inability for the radical to be formed. The concentration of trapped polystyryl radical will be directly dependent upon the amount of initiator, the more initiator is present, the more radical sites may be formed and then trapped. Regardless of the amount of initiator (di-*t*-butylperoxide) which is present, no trapped material is ever found; if radicals are produced along the polystyrene chain, there will be an observable concentration of trapped radicals present in the sample. The mobility of the initiator radical will not be significantly different from that of the nitroxide radical so if the radical can be produced, it can be trapped. One can discard any special stability of the radical since it is possible to observe NMR spectra of the polystyrene after reaction. If radicals were present along the polymeric chain, these would be obvious in the NMR spectrum. This leads to the conclusion that radicals are never formed from the interaction of polystyrene and polymerization initiators.[18]

Anionic Graft Copolymerization onto Polystyrene. We have been able to initiate graft copolymerization onto polystyrene by the use of anionic initiators but even here the reaction does not proceed as one may expect. Once again one would likely think that the benzylic hydrogen is the most acidic and should be the one removed by a lithium reagent.

This is not the case and we have been able to show that it is more likely, based upon both kinetic and thermodynamic acidity arguments, that a ring hydrogen atom will be removed and this is the experimental observation. The site of lithiation has been identified by NMR spectroscopy after trapping of the polystyryl anion with either trimethylchlorosilane or iodomethane. Not all monomers are compatible with organolithium reagents; we have used acrylonitrile since our principal interest in this chemistry is to add a monomer which will give a high char yield so that the possibility of thermal protection is offered to the polymer. A typical monomer which has been used in other work from this laboratory has been methacrylic acid. This clearly is not compatible with a lithium reagent. The site at which graft copolymerization occurs has been identified by ^{13}C NMR spectroscopy on the derivatized lithiated polystyrene.[21]

Photochemical Graft Copolymerization onto Styrene-Butadiene Block Copolymer

Both photochemical procedures are only useful for polymers which contain double bonds. The anthracene procedure generates an oxygen-based radical at the position allylic to the double bond while benzophenone causes addition to the double bond. There are published infrared spectra which are suggested to support this mode of addition for the benzophenone procedure. Abdel-Razik has used benzophenone photochemistry to cause the addition of acrylamide to ABS[3] while this group has used the anthracene photosensitized formation of singlet oxygen to bring about the addition of both acrylamide and methacrylic acid to ABS.[7] Since the conditions for the two reactions are not comparable, it is difficult to compare the two results in absolute terms but it is clear that the graft yield is significantly higher for the acrylamide/benzophenone system than for the acrylamide/anthracene system. There are two explanations which may be offered for these differences: solvent effects and difference in the position and type of addition. The benzophenone photochemistry was performed in chloroform solution while the anthracene mediated graft copolymerization was performed in water solution. Chloroform will swell the polymer and thus mediate solubility within the film, and hence graft copolymerization, while water will not swell ABS. The other difference may be equally significant. For the anthracene mediated reaction, addition of monomer occurs to an oxygen-based radical at an allylic position. An oxygen radical is unlikely to be very stable; this implies that chain transfer may easily occur limiting the number of such radicals and hence the extent of graft copolymerization. On the other hand, for the benzophenone-mediated reaction a secondary carbon radical is produced and this should be more stable than the oxygen radical and thus less likely to chain transfer, giving higher graft yields.

The photochemical initiation of graft copolymerization onto SBS has only been studied with the anthracene mediated process. Very high graft yields, in excess of 1000%, may be obtained under some conditions.[2] The amount of monomer which may be attached to the polymer by graft copolymerization is dependent upon the concentrations of anthracene and monomer and the temperature of the reaction. As the concentration of anthracene is increased, the concentration of hydroperoxides also increase; the concentration of hydroperoxides also depends upon the irradiation time. Of course, if more hydroperoxides are present, more radicals may be formed which leads to more sites at which monomer may add. The temperature dependence is attributable to the necessity to break the hydroperoxide to give the oxygen radical; the higher the temperature the faster will be the rate at which the radicals are formed. In general one may state that the observations made for the ABS case are valid for the SBS system. The graft yields are much higher for SBS than for ABS. This may be due to the much greater amount of butadiene in SBS or the greater accessability of the butadiene in SBS.

Chemical Initiation of Graft Copolymerization onto Styrene-Butadiene Block Copolymer

Huang and Sundberg[22-25] have extensively studied the graft copolymerization of monomers onto polybutadiene. Since the above results for polystyrene indicate that it is not possible to graft copolymerize onto polystyrene by a radical mechanism, the graft copolymerization of monomers onto SBS should follow the identical pattern as observed for polybutadiene. Huang and Sundberg indicate that there are two processes by which one may graft onto a butadiene unit. The first of these begins by the interaction of the initiator radical with the polymer leading to the removal of a hydrogen atom and subsequent addition of monomer units to this site while the second involves the addition of a polymeric radical to the double bond of the butadiene. These are shown in Scheme 3.

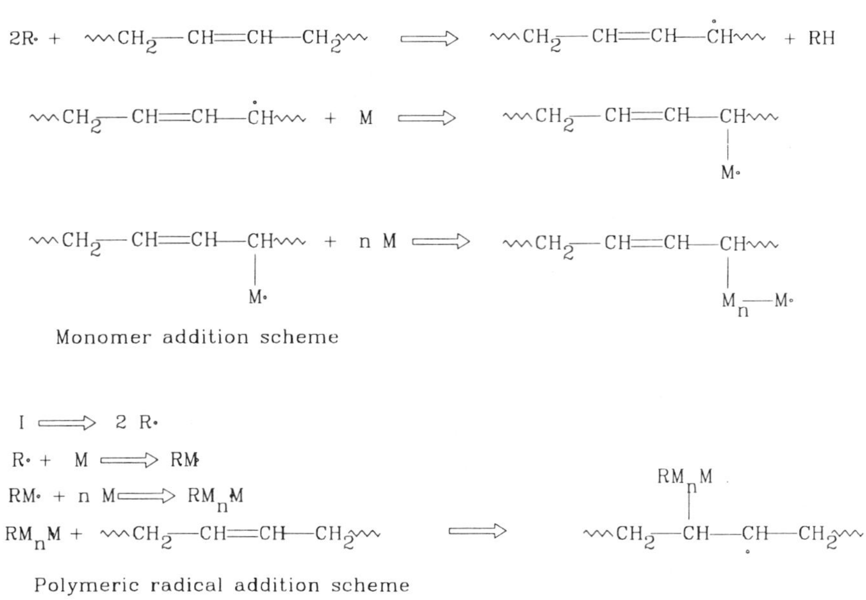

Scheme 3. Two pathways for the addition of monomers to butadiene units

The monomer addition scheme, shown at the top, requires an initiator which is capable of removing a hydrogen atom from the allylic position of the butadiene; resonance stabilization of the radical from AIBN does not permit this initiator to effect this reaction while benzoyl peroxide is capable of reaction to remove a hydrogen atom and initiate the reaction. On the other hand the polymeric radical addition scheme requires that homopolymerization of the monomer be initiated and this macroradical then attack the polymer and lead to the formation of the graft copolymer. Huang and Sundberg explain that the reactivity of the monomer relative to that of the macroradical controls which process occurs. A reactive monomer, such as a methacrylate or styrene, will react at a radical site and the addition of monomer at the allylic site occurs. A non-reactive monomer, such as an acrylate, will in turn have a more reactive macroradical and this will be the reactive species. Mayo and coworkers[26] have established this principle of reactivities of monomers and macroradicals many years ago. The reactivity of the monomer is controlled by inductive and hyperconjugative effects. Thus a methacrylate is reactive because of the inductive effect and hyperconjugation which occurs through the presence of the methyl group and its absence, in the acrylate, leads to the loss of this reactivity.

In this laboratory we have studied the interaction of methyl methacrylate, methyl acrylate, methacrylic acid, and acrylic acid with the goal of positively identifying the site of addition to the polymer and hence proving the type of reaction. A comparison of the results for both BPO and AIBN as initiators is suggestive of the actual mechanism and site of attachment but it is not conclusive proof of the actual site of attachment.

The results that have been found for these four monomers are rather surprising. For three of the monomers, methyl acrylate, acrylic acid, and methacrylic acid, both BPO and AIBN will initiate the graft copolymerization while for methyl methacrylate only BPO gives a reasonable yield of the graft copolymer. Even more surprising is the fact that for the two esters, no ungrafted SBS is found while for the acids only a small amount of the SBS is actually involved in the graft copolymerization.

The reactions have been performed in chloroform solution and the monomers of methacrylic acid and acrylic acid have good solubility but the homopolymers are not soluble and the graft copolymers are also not soluble. For the esters the homopolymers are soluble in the reaction solvent. The consequence of this is that homopolymerization and graft copolymerization of the esters occurs in a homogeneous system while reactions of the acids occur in a heterogenous system. Since both AIBN and BPO are equally effective catalysts for the reaction, it would appear that the monomer undergoes homopolymerization to give a polymeric radical and that this polymeric radical is the reactive species. The homogeneity of the reaction system alone does not explain the observation that only a small amount of polymer is grafted by the acids while all is grafted when the esters are used but it does provide useful information which may be combined with other facts to better understand the course of the reaction.

A constant fraction of the monomer is converted to graft copolymer for the reactions of methyl methacrylate and methyl acrylate with SBS; this is independent of the amount of monomer and the amount of initiator. Both homopolymer and graft copolymer are formed and 25 - 30% of the monomeric methyl methacrylate reacts to form the graft copolymer while the remainder forms homopolymer; the fraction of graft copolymer is close to 40% for methyl acrylate. This is also true for the reaction of acrylonitrile with polystyryllithium; here the amount of graft copolymer is a little lower, in the range of 15 - 20%.[18]

The graft yield for the acids is much higher than it is that for the esters. The maximum graft yield which has been observed is around 40% for methyl methacrylate and 70% for methyl acrylate; reactions conditions such as quantities of reagents and temperature can greatly effect the extent of reaction. For methacrylic acid using BPO as initiator the graft yield exceeds 200% under some conditions; for acrylic acid even higher values have been observed under conditions which are similar to those used for the esters.

If one combines all of this information, one can develop a scheme to account for all of these observations. There is a competition for initiator in the reaction of SBS with methyl methacrylate; some is used to initiate the homopolymerization of the monomer and some is used to remove an allylic hydrogen from the SBS chain. Once both of these sites are initiated, they will both compete for monomer. In the other three cases the initiator can once again initiate both the polymer and the monomer. For these monomers the reactivity of the monomer is too low and it will only react with its own polymeric radical to build up a bigger polymeric radical. The polymeric radical has two double bonds to which it may add, that of the monomer and that of the SBS. It is apparent that SBS is more reactive to the polymeric radical of methyl acrylate than it is to the polymeric radical of either methacrylic acid or acrylic acid. Thus the polymeric radical of methyl acrylate forms and eventually it reacts with SBS to give the graft copolymer but everything remains in solution. SBS is less reactive than is monomer towards the polymeric radical of both acids; this means that rather large polymeric radicals can build up until the concentration of monomer decreases to such a point where the concentration of the SBS makes it competitive with monomer. Now as soon as the polymeric radical reacts with SBS, the graft copolymer will precipitate from the solution so most of the monomer is relatively rapidly consumed in the formation of homopolymer and graft copolymer which both precipitate from the solution.

Site of Grafting

The site on the SBS at which graft copolymerization has occurred has been identified by infrared spectroscopy. Three bands are of interest: the olefinic C-H stretching frequency near 3060 cm^{-1}, the CH_2 bending mode in the region of 1450 cm^{-1}, and the olefinic C-H out-of-plane bending mode near 910 cm^{-1}. If reaction occurs at an allylic site, this will change the intensity of the CH_2 bending mode but have little, if any, effect on the olefinic modes. On the other hand, if reaction occurs at the double bond, one expects that the two olefinic modes will decrease while the CH_2 mode will be little effected. The intensities of these bands must be normalized and this has been accomplished by taking ratios of the intensities at the various positions. It is surprising that a difference can be seen in the infrared spectrum. It might be expected that the number of positions along the SBS chain which are modified would be very small and that differences in absorbance would not be seen. The results for methyl methacrylate and methyl acrylate are shown in Table 1; these show that there are differences in the absorbances.

Table 1. Infrared Absorbances for Selected Carbon-Hydrogen Bending Modes

System	Absorbance at 3060/1450	Absorbance at 910/1450
SBS	0.80	0.58
MMA/BPO	0.82	0.65
MMA/AIBN	0.78	0.56
MA/BPO	0.60	0.51
MA/AIBN	0.60	0.38

One sees that there is very little change in the ratio of the olefinic C-H stretching frequency to that of the CH_2 bending frequency for methyl methacrylate but a significant change for methyl acrylate. Likewise there is much more of a change in the ratio of the absorbance at 910 cm^{-1} to that at 1450 cm^{-1} for methyl acrylate than for methyl methacrylate. For the system methyl methacrylate with benzoyl peroxide one sees a small increase in both ratios. These observations, decrease in olefinic intensities for methyl acrylate but not for methyl methacrylate, are completely consistent with methyl methacrylate adding at an allylic site but methyl acrylate adding to the double bond.

The infrared spectra for the acids are much more difficult to interpret. The acids contain a hydroxyl functionality which absorbs in the range of 3500 cm^{-1}. Unfortunately when a large amount of acid is present this band extends down into the C-H stretching frequency region and completely obscures all C-H stretching vibrations. The CH_2 bending mode is also now not a distinct band but rather is one peak among many in this region and the resolution is not sufficient to measure an intensity. This leaves only the C-H out-of-plane bending mode for identification. The spectra which can be obtained are quite similar to those for the graft copolymerization of acrylic acid onto acrylonitrile-butadiene-styrene terpolymer in which it has been shown that reaction occurs by addition to the double bond.[9] Since this is also completely consistent with the observations with respect to initiator efficiency, it is believed that both of the acids react at the double bond site.

Because of the observation that methyl methacrylate reacts *via* the monomer while methyl acrylate reacts through the polymeric radical, the expectation when this work was begun was that methacrylic acid would add at an allylic site while acrylic acid would add at the double bond. Since this proves not to be the case, an explanation is required. Mayo's ordering[26] of the reactivity of monomers is based upon the electron donating ability of the substituents around the double bond. The hyperconjugative and resonance effects of the methyl group render methyl methacrylate more reactive than is methyl acrylate. The presence of the methyl substituent will enhance the reactivity of methacrylic acid relative to that of acrylic acid but the replacement of the methoxy by a hydroxy lowers the electron donating ability and decreases the reactivity of the monomeric acids. This is apparently the deciding factor and the acids can only react through the polymeric radical.

CONCLUSION

One may perform radical graft copolymerizations onto the butadiene region of copolymers of styrene and butadiene without any reaction occurring at the styrene portions of the copolymer. If the monomer is reactive, reaction of the monomer at an allylic site occurs while for less reactive monomers, the polymeric radical is formed and this adds to the double bond of the polymer. Proof of the site of grafting comes from information about the relative efficiency of different initiators but the most important information is obtained from infrared spectroscopy. One can observe differences in the spectra which can be related to the mode of addition.

REFERENCES

1. G. Geuskens and Ph. Thiriaux, Surface modification of polymers. II. Photo-oxidation of SBS containing anthracene and grafting initiated by photo-induced hydroperoxides, *Eur. Polym. J.* 29: 351 (1993).
2. X. Dong, G. Geuskens, and C. A. Wilkie, Graft copolymers of methacrylic acid and SBS and K-Resin by the anthracene photosensitized formation of hydroperoxides, *Eur. Polym.* J., 31: 1165 (1995).
3. E. A. Abdel-Razik, Photoinduced graft copolymerization of acrylamide onto styrene-butadiene-acrylonitrile copolymer, *J. Photochem. Photobiol. A: Chem*, 69:121 (1992).
4. D. Ruckert, G. Geuskens, P. Fondu, and S. Van Erum, Surface modification of polymers - III. Photo-initiated grafting of water soluble vinyl monomers and influence on fibrogen adsorption, *Eur. Polym. J.*, 31:431 (1995).
5. J. P. Kennedy, Ed., *Cationic Graft Copolymerization*, Wiley, New York, 1977.
6. C. A. Wilkie, M. Suzuki, X. Dong, C. Deacon, J. A. Chandrasiri, and T. J. Xue, Grafting to achieve flame retardancy, *Polym. Degrad. Stab.*, 54: 117 (1996).
7. M. Suzuki and C. A. Wilkie, Graft copolymerization of methacrylic acid and acrylamide onto acrylonitrile-butadiene-styrene terpolymer by photoinduced hydroperoxide, *J. Polym. Sci.: Part A: Polym. Chem.*, 33: 1025 (1995).

8. C. Deacon and C. A. Wilkie, Graft copolymerization of acrylic acid onto acrylonitrile-butadiene-styrene terpolymer and thermal analysis of the copolymers, *Eur. Polym. J.*, 32: 451 (1996).
9. J. A. Chandrasiri and C. A. Wilkie, Chemically initiated graft copolymerization of acrylic acid onto acrylonitrile-butadiene-styrene (ABS) terpolymer and its constituent parts, *J. Polym. Sci: Part A: Polym. Chem.*, 34: 1113 (1996).
10. E. Uchida, Y. Uyama, and Y. Ikada, Surface graft polymerization of acrylamide onto poly(ethylene terephthalate) film by UV irradiation, *J. Polym. Sci.: Part A: Polym. Chem.*, 27: 527 (1989).
11. Y. Uyama and Y. Ikada, Electrostatic properties of UV-irradiated and surface-grafted polymers, *J. Appl. Polym. Sci.*, 41:619 (1990).
12. K. Allmer, A. Hult, and B. Ranby, Surface modification of polymers. I. Vapour phase photografting with acrylic acid, *J. Polym. Sci.: Part A: Polym. Chem.*, 26: 2099 (1988).
13. K. Allmer, A. Hult, and B. Ranby, Surface modification of polymers. II. Grafting with glycidyl acrylates and the reactions of the grafted surfaces with amines, *J. Polym. Sci.: Part A: Polym. Chem.*, 27: 1641 (1989).
14. K. Allmer, A. Hult, and B. Ranby, Surface modification of polymers. III. Grafting of stabilizers onto polymer films, *J. Polym. Sci.: Part A: Polym. Chem.*, 27: 3405 (1989).
15. K. Allmer, A. Hult, and B. Ranby, Surface modification of polymer films. IV. UV initiated degradation of polymers with stabilizers grafted onto the surface, *J. Polym. Sci.: Part A: Polym. Chem.*, 27: 3419 (1989).
16. J. Xue and C. A. Wilkie, Swelling assisted modification of poly(ethylene terephthalate) by methacrylic acid, *J. Polym. Sci.: Part A: Polym. Chem.*, 33: 1019 (1995).
17. T. J. Xue and C. A. Wilkie, The interaction of vinyl monomers and poly(ethylene terephthalate) in the presence of various initiators produces a physical mixture not a graft copolymer, *J. Polym. Sci.: Part A: Polym. Chem.*, 33: 2753 (1995).
18. T. J. Xue and C. A. Wilkie, Graft copolymerization of acrylonitrile onto polystyrene, *J. Polym. Sci.: Part A: Polym.Chem. Ed.*, in press.
19. D. D. Jiang and C. A. Wilkie, Chemical initiation of graft copolymerization of methyl methacrylate onto styrene-butadiene block copolymer, *J. Polym. Sci.: Part A: Polym. Chem Ed.*, in press.
20. D. D. Jiang and C. A. Wilkie, Graft copolymerization of methyl acrylate, methacrylic acid and acrylic acid onto styrene-butadiene block copolymer, submitted.
21. T. J. Xue, M. S. Jones, J. R. Ebdon, and C, A. Wilkie, Lithiation - alkylation of polystyrene occurs only on the ring, *J. Polym. Sci.: Part A: Polym. Chem Ed.*, 35: 509 (1997).
22. N-J. Huang and D. C. Sundberg, Fundamental studies of grafting reactions in free radical copolymerization. I. A detailed kinetic model for solution polymerization, *J. Polym. Sci.: Part A: Polym. Chem.*, 33: 2533 (1995).
23. N-J. Huang and D. C. Sundberg, Fundamental studies of grafting reactions in free radical copolymerization. II. Grafting of styrene, acrylate, and methacrylate monomers onto *cis*-polybutadiene using AIBN initiator in solution polymerization, *J. Polym. Sci.: Part A: Polym. Chem.*, 33: 2551 (1995).
24. N-J. Huang and D. C. Sundberg, Fundamental studies of grafting reactions in free radical copolymerization. III. Grafting of styrene, acrylate, and methacrylate monomers onto *cis*-polybutadiene using benzoyl peroxide initiator in solution polymerization, *J. Polym. Sci.: Part A: Polym. Chem.*, **33**, 2571 (1995).
25. N-J. Huang and D. C. Sundberg, Fundamental studies of grafting reactions in free radical copolymerization. VI. Grafting of styrene, acrylate, and methacrylate monomers onto vinyl-polybutadiene using benzoyl peroxide and AIBN initiators in solution polymerization, *J. Polym. Sci.: Part A: Polym. Chem.*, 33: 2587 (1995).
26. F. R. Mayo, F. M. Lewis, and C. Walling, Copolymerization. VIII. The relationship between structure and reactivity of monomers in copolymerization, *J. Am. Chem. Soc.*, 70: 1529 (1948).

CHEMICAL CROSS-LINKING BY GLUTARALDEHYDE BETWEEN AMINO GROUPS: ITS MECHANISM AND EFFECTS

Jun-ichi Kawahara,* Keiichiro Ishikawa, Tadafumi Uchimaru and Haruo Takaya

National Institute of Materials and Chemical Research
Higashi, Tsukuba, Ibaraki 305, JAPAN

INTRODUCTION

Chemical cross-linking reaction, although rather simple as a principle, is a powerful tool for processing polymeric materials, since it can significantly alter the molecular structure of materials as a whole, even if the alteration of the structural unit might be little. This situation --- the properties of cross-linked materials could significantly differ from those of the original polymers --- could be compared to the fact that the properties of polymers significantly differ from those of their monomers.

Chemical cross-linking reaction by glutaraldehyde (GA)** between amino groups would be representative for this purpose. It is now considered to be quite useful in constructing functional polymeric materials from proteins, aminopolysaccharides, and other natural and synthetic polymers that contain amino groups, besides the wide range of application in life science, probably represented by tissue fixation.[1] It is also considered as a possible candidate to replace the current tanning technology that uses chromium or formaldehyde, which are hazardous to the environment and health.

Since the development of these methodologies would rely heavily upon an understanding of the chemistry of GA and its cross-linking mechanism, it would be a matter of course that there have been a number of studies on them. Nevertheless, there still remains a considerable problem in the fact that most studies[2-5] neglected the possible solvent effects on the GA structure. For example, Monsan et al., upon whose study[5] current theory of cross-linking mechanism is based,[6, 7] made analyses of GA structure only in organic solvents which are typical in the analytical methods they used, although commercial GA is supplied in, and the cross-linking reaction of GA is carried

*To whom correspondence should be addressed.
**Abbreviations used: GA, glutaraldehyde; ESI-MS, electrospray ionization mass spectrometry

Polymer Modification
Edited by Swift *et al.*, Plenum Press, New York, 1997

out in aqueous solution, and GA could react with water in various ways as described later. Furthermore, in anhydrous solvents the equilibrium between monomeric and polymerized GA possibly shifts to the latter, which produces water.[8] Some researchers made experiments in D_2O.[2, 3] However, since deuterium exchange could occur at α - carbon,[9] it might give erroneous results to compare the peak intensities of ^{1}H- NMR. Furthermore, the hydration equilibrium constants for monoaldehydes are reported to differ between H_2O and D_2O,[10] and this will probably be the case also with GA.

Besides, although some researchers measured ^{13}C- NMR,[4] it is known that comparison of the peak intensities is not quantitative in simple ^{13}C- NMR.[11]

In this article, the chemistry of GA was investigated using UV spectroscopy, light scattering, ^{13}C- NMR and electrospray ionization mass spectrometry (ESI- MS). In these analyses, we used solely natural water in order to avoid the above- mentioned problems, and also paid attention to the quantitativeness of the analyses. It was found that the actual cross- linking mechanism of GA is quite unique; it is largely different from that in the current theory in the field of biochemistry[6, 7] and also from that of most other cross- linking reagents. Finally, the expected effects of this cross- linking reaction on the properties of the resultant materials are also discussed.

EXPERIMENTAL

As GA sample, commercial 70% GA in aqueous solution (EM grade, purchased from Wako in ampules, pH3.5- 4.0) was used without purification. The sample was diluted to the indicated concentration using redistilled natural water purged of the dissolved oxygen by gaseous nitrogen, and if necessary, phosphate buffer was added more than several hours later.[8]

UV- VIS absorption was measured with Beckman DU- 70 spectrophotometer, using cuvettes of 10mm or shorter pathlength. Light scattering was measured with Otsuka

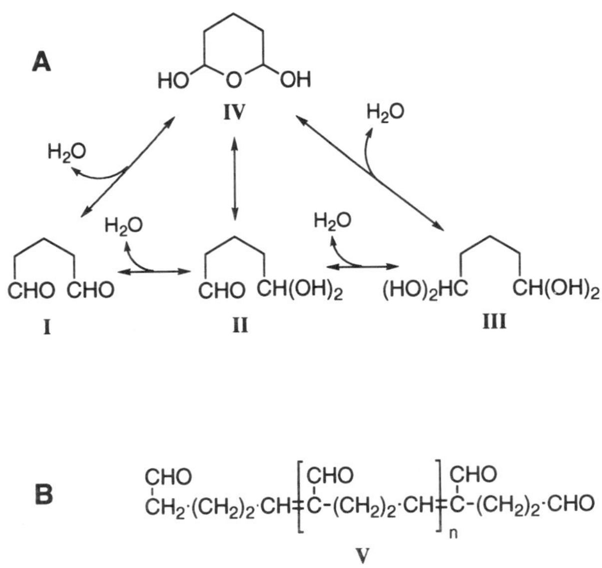

Figure 1. A: The molecular structure of GA in non- concentrated aqueous solution and the possible conversion paths. **B:** The principal structure of GA in aqueous solution in the current theory in biochemistry.

DLS-700S light scattering photometer at 633nm, calibrated with benzene. The optical clarification was performed with teflon filters. The specific refractive index increment (dn/dc) was obtained with Chromatix KMX-16 refractometer at the same wavelength, calibrated with NaCl solution.

^{13}C-NMR was measured in the mode of ^{1}H noise decoupling without NOE, using D_2O sealed up in a capillary tube. All UV and NMR measurements were performed under precise temperature control.

ESI-Mass spectra were obtained using JEOL HX/HX110A mass spectrometer equipped with ESI source (Analytica of Branford, Branford, CT) and channel array detector, calibrated in the FAB mode with alkali metal iodides. Sample solutions were diluted 2000-fold with 50% aqueous methanol (v/v) containing 0.1% acetic acid.

RESULTS AND DISCUSSION

Structure of GA

The molecular structure of GA in aqueous solution was investigated, using UV-VIS spectroscopy, light scattering and ^{13}C-NMR.[8, 12] It was found that GA is exclusively monomeric --- mainly cyclic hemiacetal (IV), accompanied with a small amount of dehydrate (I), monohydrate (II), and dihydrate (III) structures, all of which are in fast equilibrium with each other --- in non-concentrated (namely, ca.10%* or less, at room temperature) GA solution (Figure 1), which is the usual condition adopted in the cross-linking reaction. It would be worthy of notice that the content of α, β-unsaturated structure (V) is negligible in any conditions (pH2-8), which structure has been believed, by many biochemists, to be the principal species in GA solution and furthermore to play the central role in the cross-linking reaction.[5, 6, 7]

Cross-linking Mechanism of GA

In the previous section, GA was shown to be exclusively monomeric at the starting point of cross-linking reaction. Our results, however, should not necessarily be interpreted that GA cross-links two amino groups in this simple monomeric form (Figure 2). Although some researchers of polymer science seem to think so, and such a fashion --- only the functional groups on the reagent side react with the functional groups on the polymer side, with no significant alteration to the molecular skeleton of the reagent --- is the case with most other cross-linking reagents, this simple mechanism is highly doubtful in case of GA, if one considers the following facts.

1. The formed cross-link with lysine (a structural unit that provides amino group in case of proteins) is stable against acid hydrolysis (110℃ in 6N HCl for 24hr).[6]
2. Intermolecular cross-linking efficiency is remarkably high, compared with other bi-functional cross-linking reagents (data not shown).

The functional group that directly participates in the cross-linking reaction would

*Below this GA concentration, the relative content of each monomeric structure was found to be virtually constant regardless of the GA concentration, and also pH (pH2-8), in contrast to the temperature which has marked effect on it. On the other hand, above this concentration, considerable amount of polymerized structures exist in GA solution, which are closely related with cyclic hemiacetal structure, that is, again not related with α, β-unsaturated structure. The conversion velocity between the monomeric and polymerized structures is quite slow, compared to that between the monomeric structures. The chemistry of GA in concentrated aqueous solution will be discussed in detail elsewhere.

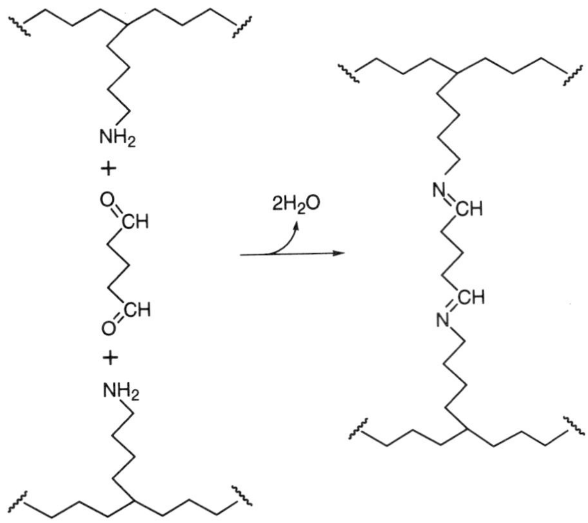

Figure 2. The cross- linking mechanism of GA? Such a fashion is the case with most other cross- linking reagents.

be the dehydrate aldehyde group. Although the amount of dehydrate form is rather limited (10- 15% of all possible forms) at room temperature,[12] this nature would not cause major hindrance to the cross- linking reaction, since the aldehyde group of all monomeric forms are in fast equilibrium with each other as mentioned above. The portion of dehydrate form consumed in the cross- linking reaction would be compensated swiftly by the dehydration of other forms.

First, the process of gel formation from aqueous solution of proteins was observed in situ, by UV- VIS spectroscopy, in order to follow up the chemical reactions on the participating molecules. It was shown that increase of UV absorption occurs during the gelation process, with the peak position shorter than 240nm (data not shown). This absorption increase was sharper, as pH increased. Since the wavelength range shorter than ca.240nm was difficult to observe due to the strong absorption of amide group, the analogues of lysine side chain, namely low molecular weight amines, were used next.

It was found that, if a small amount of n- amylamine whose molecular structure is just the same as that of lysine side chain is added to GA solution, there is swiftly produced a significant absorption peak near 235nm (Figure 3). This peak closely resembles the one in aqueous solution expected for the π - π^* allowed transition of the ethylenic double bond that is in conjugation with an adjacent double bond (R_1 - C=C(R_2) - C=N or R_3 - C=C(R_4)- C=O).[12, 13] Such a structure is only produced when GA itself is polymerized through aldol condensation, followed by dehydration. It is assumed that the reaction mechanism of GA with proteins and other polymers containing amino group is basically similar to that with simple amines, judging from the resemblance of spectral change in the longer wavelength range and pH dependency of the reaction.

The polymerization reaction of GA was further confirmed by the reaction of ethanolamine with GA observed by light scattering (Figure 4). This amine was highly soluble even after the reaction with GA,* and therefore, quite useful for in situ observation and also for the analyses by ESI- MS, which are described later.

---

*The reaction products between GA and amines have a tendency to be less soluble in aqueous solution, forming aggregates if the sample is not so dilute. Sample aggregation hinders most analyses, including UV spectroscopy, light scattering, and ESI- MS. The cause of solubility decrease is discussed in some detail later, at the section "The Effects".

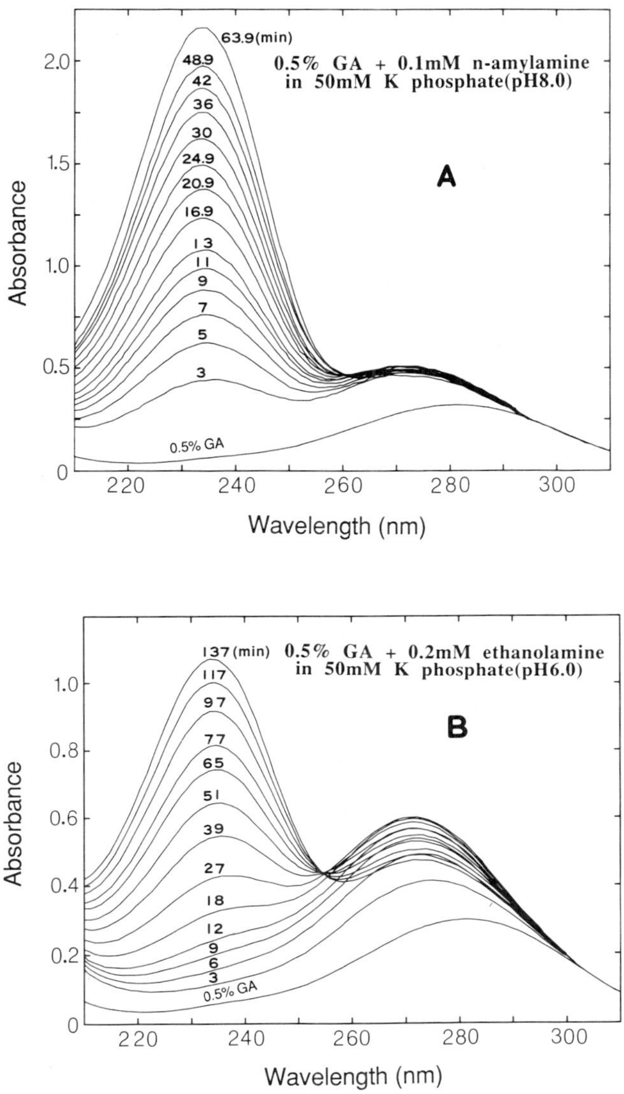

Figure 3. Spectral change of GA induced by amine. Numbers in the figure indicate the time in minutes after the addition of amine to GA solution.

Next, the effect of the amine structure was investigated by UV spectroscopy (Figure 5). It was found that only primary amines induce the polymerization of GA, regardless of their basicity or reaction pH. This fact strongly suggests that aldol condensation of GA is not induced by the base catalysis of the lone pair of amines, but that the formation of Schiff base imine between one monomeric GA and one amino group is the starting step of the following condensation polymerization (Figure 6). In this reaction scheme, conversion of intermediate structure VI to VIII via VII would be facili-

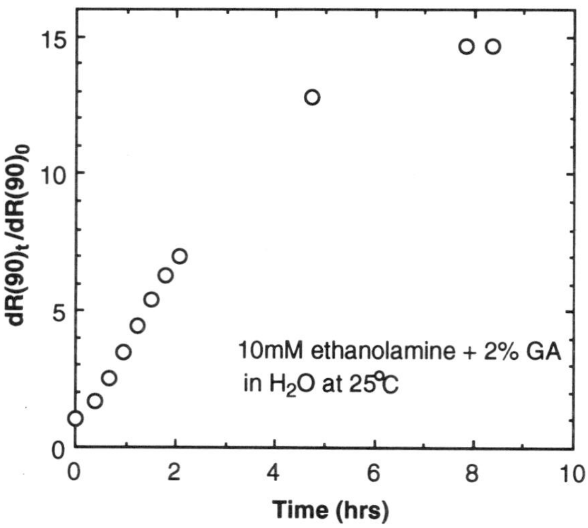

Figure 4. Polymerization of GA induced by amine observed by light scattering. The normalized scattering intensity at a right angle, which is virtually equivalent to the average polymerization degree of GA under the conditions used here, is plotted against the time after the addition of amine to GA solution.

tated by nitrogen atom substituted for carbonyl oxygen.

In Figure 3, there appears a minor absorption peak on the side of longer wavelength. The appearance of this peak was always temporary, and its intensity and duration varied significantly according to pH, structure of amine, and also the concentration of amine and GA. In contrast, by taking the first derivative of the spectra ($dA/d\lambda$) it was shown that the peak position is actually constant all through the reaction time course (λ_{max} = ca. 270nm), and this λ_{max} is virtually the same regardless of the

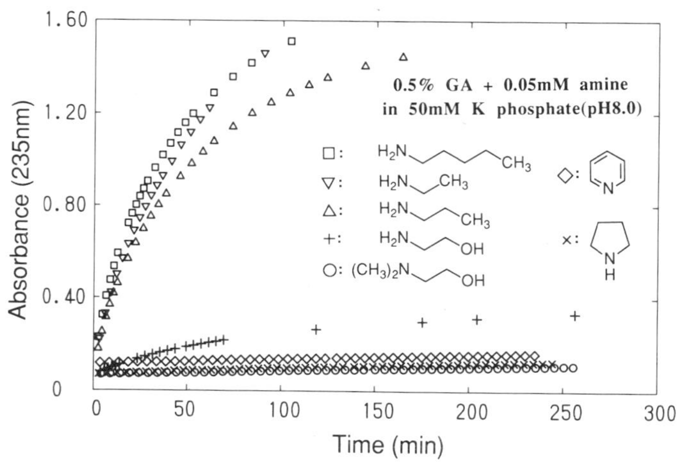

Figure 5. Polymerization of GA induced by amine observed by UV absorption. Each amine was added to GA solution at 0 min.

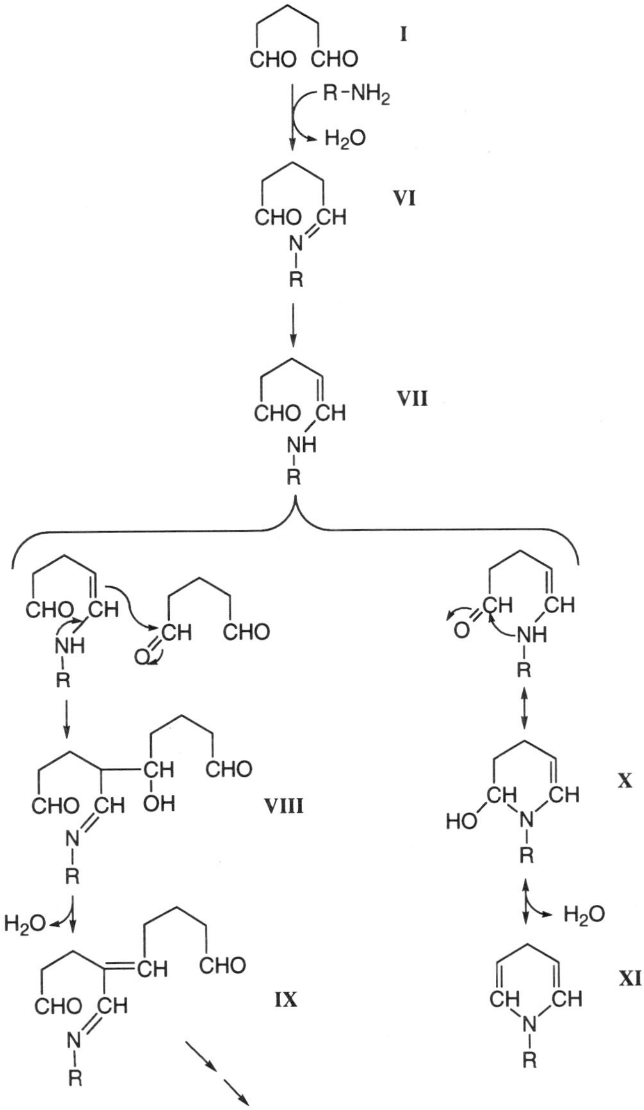

Figure 6. Possible cross- linking mechanism of GA.

amine structure and other conditions used (data not shown), just as the case with the main peak (λ_{max} = ca. 235nm).

On the other hand, one of the intermediate structures, namely structure VII, seems to be possibly converted to 1,4- dihydropyridine structure (XI), from the structural point of view. Besides, since λ_{max} of band III[14] for XI is expected to be similar to that of the above side peak,[15] the appearance of this peak would indicate the actual formation of XI.* From the observation that the appearance of this peak was always temporary, and also from the structural relation between XI and the end structure IX - - - the expected

*In Figure 3A, in contrast to Figure 3B, some portion of the "main peak" would be ascribed to band I[14] of deprotonated form of XI.[15] Nevertheless, this would not affect at all the former discussion - - - the swift formation of ethylenic double bond that is conjugated with an adjacent double bond. This deduction is clearly confirmed by Figure 3A; "main peak" continues to increase sharply even after the "side peak" begins to decrease.

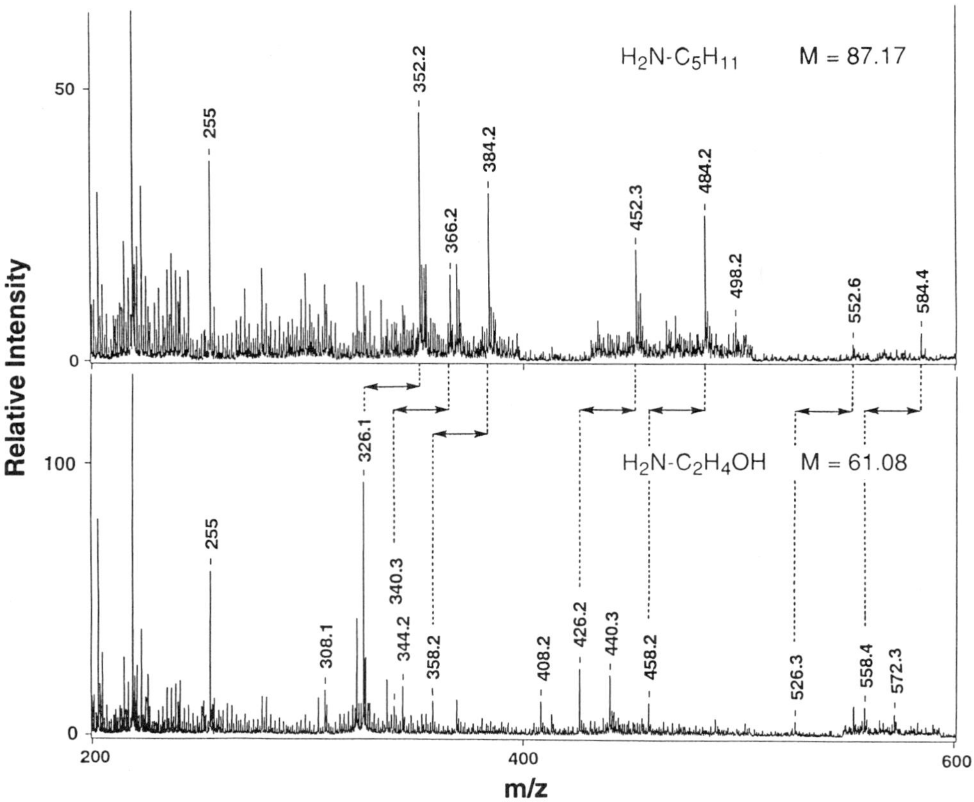

Figure 7. ESI- MS spectra of GA- amine products. 1 mM of each amine was added to 0.5%GA in H_2O and left for 2 days at 25℃ before analysis.

end structure of that GA monomer unit, which was confirmed by ESI- MS, as described later - - - , it is deduced that XI would not locate along the main path of the reaction, but at the terminal of the reversible side path branching off from VII (Figure 6). Figure 3 and other data obtained under different conditions indicate that the conversion between VII and XI is much faster than the reaction from VII to IX along the main path, in usual conditions.

In this reaction scheme, the quantity and the duration of XI would depend on the velocity of the formation and consumption of VII, and also on the velocity of forward and reverse reactions between the structures VII and XI, thus accounting for the observation that they are significantly affected by the reaction conditions.

Finally, the aqueous solution of reaction products between GA and amines was directly ionized by electrospray ionization, and subsequently analyzed by mass spectrometry (Figure 7). The upper half is the result between GA and amylamine and the lower is the one between GA and ethanolamine. Comparing the two results, it is shown that mass difference of the corresponding signals, linked by arrows in the figure, is just equal to 26. Since the molecular weight difference between these two amines is 26, this result indicates that amine molecule was actually incorporated covalently into GA oligomer, and the number of amine molecules incorporated per one GA oligomer is just one, in this condition. Assuming that one amine molecule is incorporated into one GA oligomer, the expected mass of GA- amine products can be calculated (Table 1).

The signals of the expected mass were actually observed using the higher concentration of ethanolamine. The observed intensities are summarized in Table 1. There was found a significant tendency in these intensities, and this tendency suggests that the dehydration step following aldol condensation occurs almost completely at GA monomer

Table 1. Expected mass[1] of GA- amine products and their observed intensities[2]

degree of polymerization[3]	degree of dehydration[4]	M[5]	observed intensity[6] (relative)
2	0	244, 245	N[7]
	1	226, 227	45
3	0	344, 345	23
	1	326, 327	166
	2	308, 309	48
4	0	444, 445	7.2
	1	426, 427	56
	2	408, 409	28
	3	390, 391	4.2
5	0	544, 545	2.1
	1	526, 527	19
	2	508, 509	14
	3	490, 491	3.9
	4	472, 473	2.5
6	0	644, 645	N
	1	626, 627	4.7
	2	608, 609	5.5
	3	590, 591	2.7
	4	572, 573	3.1
	5	554, 555	N
7	0	744, 745	N
	1	726, 727	1.9
	2	708, 709	1.7
	3	690, 691	1.1
	4	672, 673	1.4
	5	654, 655	N
	6	636, 637	N

[1] Assuming just 1 molecule of ethanolamine is incorporated into each GA oligomer.
[2] 10mM ethanolamine was used, with all the other conditions the same as Figure 7.
[3] Number of GA monomers incorporated into that GA oligomer. See Figure 8.
[4] Number of dehydration steps occurred in that GA oligomer. See Figure 8.
[5] Mass of monoprotonated product molecules which contain no or one ^{13}C atom.
[6] Summation of the 2 peak heights in the left adjacent column.
[7] Negligible.

units containing Schiff base imine, in contrast to the GA units containing no Schiff base, where little dehydration occurs (Figure 8). Therefore, the formed Schiff base linkage eventually constitutes a conjugate system with the adjacent ethylenic double bond. Once such conjugation is formed, the resonance interaction is reported to give Schiff base linkage the stability to acid hydrolysis.[5]

The possibility of solvent association with carbonyl group on the molecular ion, whose polarity is stronger than imino group, is eliminated, since solvent association with

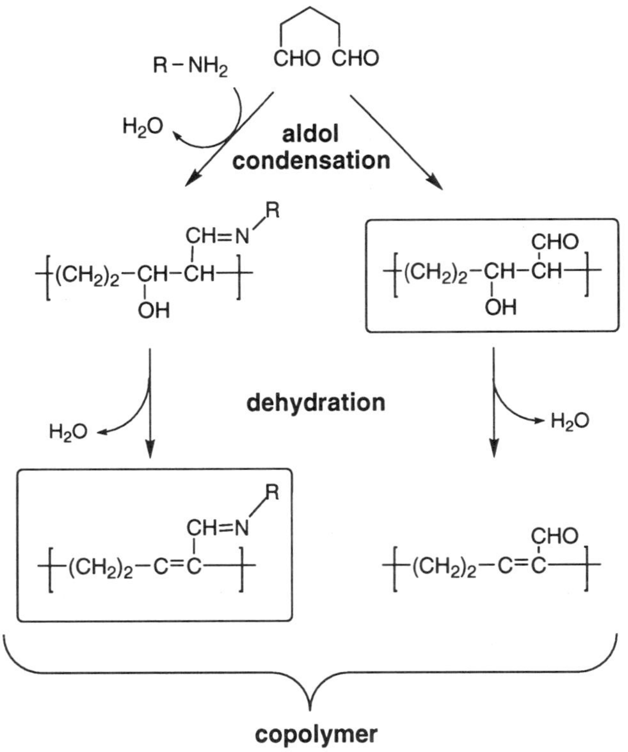

Figure 8. Deduced reaction scheme in the formation of GA cross-links. Predominant final structural units are enclosed by rectangles.

amide group, whose polarity is even stronger than simple carbonyl group, is negligible in the ionization conditions used here.[16] Therefore, the observation that dehydration occurs little at locations adjacent to carbonyl group would reflect the true structure, not the artefact caused by the association of one water molecule per one carbonyl group, whose neighboring α, β-position actually dehydrated.

The final structure of the formed cross-link proposed in this article might seem similar to that depicted in the current theory in the aspect that it is not monomeric but polymerized GA, although some differences surely exit; for example, unsaturation at α, β-position occurs only at locations adjacent to Schiff base imine according to our theory. The most important difference between the two theories would be the timing when GA polymerization occurs. This difference would be crucial when we attempt to control the structure --- for example, the chain length --- of the formed cross-links in the future.

The Effects

When materials are processed using the cross-linking reaction by GA, and especially when the construction of new functions is attempted, the effects of this chemical reaction on the properties of the resultant materials should be kept in mind. Based upon the cross-linking mechanism discussed above, the following effects are expected.

1. Limitation of molecular chain movement

   Although the distance between cross-linking points might not be so limited, since the

Table 2. Basicity of the nitrogen lone pair.

Formula	type of hybridization	example	pKa of conjugate acid[1]	Ref
>N−	sp^3	monoalkylamine	10.57- 64 (25℃)	17, 18
=N−	sp^2	pyridine[2]	5.17 (25℃)	19
≡N	sp	acetonitrile	virtually no basicity in aqueous solution	

[1] Value for the substance exemplified in the left adjacent column.
[2] Imino double bond is in conjugation with ethylenic double bonds.

cross- linking chain cannot extend infinitely following the motion of the main chain either, the formation of cross- links would surely cause some limitation on the main chain mobility of constituent molecules of the material.

2. Decrease of the hydrophilicity
   Solubility of the resultant materials in aqueous solution tends to decrease, probably due to the following causes.

- Decrease of entropy advantage accompanying dissolution
  (inevitable thermodynamic effect of cross- linking reaction and polymerization of GA)
- Net disappearance of polar group (namely, carbonyl group)
- Dehydration at the location adjacent to formed Schiff base (disappearance of another polar group)
- Possible loss of electric charge at nitrogen atom near neutral pH, due to the cause discussed below

The following is valid, only when GA is reacted with low molecular weight amines.

- Polarity of the formed imino group is weaker than the original carbonyl group
- Addition of hydrophobic group (namely, alkyl group from amines)

3. Decrease of the basicity
   The basicity of the formed imino group would be weak significantly compared to the original amino group, due to the alteration of the hybridization type of the nitrogen lone pair.* The basicity of the nitrogen lone pair decreases as the contribution of s orbital to its hybrid orbital increases (Table 2). The expected basicity of the formed imino group would be similar to that of pyridine, whose hybridization type is the same, and whose imino double bond is conjugated with ethylenic double(s) likewise. This marked basicity decrease could cause, for example, considerable variation in the character of pH- responsive materials.**

CONCLUSION

The chemical cross- linking reaction by GA between amino groups, which has wide

*Lone pair constitutes a hybrid orbital together with sigma bond(s).
**This description combined with Table 2, however, does not necessarily mean that the actual variation of pKa of the relevant group in the material is from ca.10 to ca.5. The actual pKa of the ionic groups in materials could be altered significantly due to the electrostatic interaction between these groups. This electrostatic interaction, on the other hand, could vary significantly according to the conformation that constituent macromolecules take, for example, the secondary structure of poly(amino acid).

range of application, was investigated using UV spectroscopy, light scattering, ^{13}C-NMR and ESI-MS, paying attention to the possible solvent effects on GA chemistry and the quantitativeness of the analyses. It was found that the actual cross-linking mechanism of GA is quite unique, largely different from that in the current theory and also from that of most other cross-linking reagents. Principal points are summarized below.

1. GA is exclusively monomeric --- a mixture of several monomeric structures which are in fast equilibrium with each other --- at the starting point of the cross-linking reaction. The content of α,β-unsaturated polymeric structure is quite negligible, which has been believed, by many biochemists, to be the principal species in GA solution and to play the central role in the cross-linking reaction.
2. Actual cross-linking reaction is not based on the simple mechanism that Schiff base linkages are formed at both ends of monomeric GA. Instead, the polymerization of GA, via aldol condensation, proceeds in parallel with the cross-linking reaction.
3. It was suggested that the formation of Schiff base imine by one GA molecule with one amino group enhances an aldol condensation of it with other GA molecules. The final cross-linking structure would be a linear aldol-condensed oligomer of glutaraldehyde, with several Schiff base linkages branching off from that.
4. It was suggested that dehydration proceeds almost completely at the GA monomer unit that contains Schiff base imine, but proceeds only little at other GA monomer units. Thus, the formed imino bond is conjugated with the ethylenic double bond, which interaction leading to its stability.
5. One of the reaction intermediates seems to be swiftly and reversibly converted to 1,4-dihydropyridine structure via a side path branching off from the main path.
6. Chemical cross-linking reaction by GA could alter some properties of the material, besides the physical network formation between the constituent macromolecules.

Further, it seems that, this unique character of GA --- exclusively monomeric originally, but forms polymeric chain in the cross-linking reaction --- accounts for the practical efficiency of GA as a cross-linking tool. That is, swift diffusion into the object material as a small molecule, and effective cross-linking even between distant points as a long chain.

ACKNOWLEDGMENT

The authors thank Dr. Masao Shimizu of their Institute for his helpful discussion.

REFERENCES

1. See, for example: N.B. Rewcastle, Glutaric acid dialdehyde; a routine fixative for central nervous system electron microscopy, *Nature* 205:207 (1965).
2. F.M. Richards and J.R. Knowles, Glutaraldehyde as a protein cross-linking reagent, *J. Mol. Biol.* 37:231 (1968).
3. P.M. Hardy, A.C. Nicholls, and H.N. Rydon, The nature of glutaraldehyde in aqueous solution, *Chem. Commun.* 565 (1969).
4. E.B. Whipple and M. Ruta, Structure of aqueous glutaraldehyde, *J. Org. Chem.* 39:1666 (1974).
5. P. Monsan, G. Puzo, and H. Mazarguil, Étude du mécanisme d'établissement des liaisons glutaraldéhyde-protéines, *Biochimie* 57:1281 (1975).
6. K. Peters and F.M. Richards, Chemical cross-linking: reagents and problems in studies of membrane structure, *Ann. Rev. Biochem.* 46:523 (1977)

7. S.S. Wong. *Chemistry of Protein Conjugation and Cross- Linking*, 101, CRC Press, Boca Raton (1993).
8. J. Kawahara, T. Ohmori, T. Ohkubo, S. Hattori, and M. Kawamura, The structure of glutaraldehyde in aqueous solution determined by ultraviolet absorption and light scattering, *Anal. Biochem.* 201:94 (1992)
9. A.H. Korn, S.H. Feairheller, and E.M. Filachione, Glutaraldehyde: nature of the reagent, *J. Mol. Biol.* 65:525 (1972)
10. L.C. Gruen and P.T. McTigue, Hydration equilibria of aliphatic aldehydes in H_2O and D_2O, *J. Chem. Soc.* 5217 (1963)
11. R.M. Silverstein, G.C. Bassler, and T.C. Morrill, *Spectrometric Identification of Organic Compounds*, 4th ed., 249, John Wiley & Sons, New York (1981).
12. J. Kawahara, Y. Nagawa, H. Ichijo, and O. Hirasa, The structure and some properties of glutaraldehyde in non- concentrated aqueous solution determined by ^{13}C- NMR and Ultraviolet Spectroscopy, *submitted.*
13. L.M. Fieser and M. Fieser, *Natural Products Related to Phenanthrene*, 184, Reinhold, New York (1949)
14. U. Eisner and J. Kuthan, The chemistry of dihydropyridines, *Chem. Rev.* 72:1 (1972)
15. E.M. Kosower and T.S. Sorensen, The synthesis and properties of some simple 1,4- dihydropyridines, *J. Org. Chem.* 27:3764 (1962)
16. R.D. Smith, J.A. Loo, R.R. Ogorzalek Loo, M. Busman, and H.R. Udseth, Principles and practice of electrospray ionization- mass spectrometry for large polypeptides and proteins, *Mass Spectrom. Rev.* 10:359 (1991).
17. R.M.C. Dawson, D.C. Elliott, and W.H. Elliott, ed., *Data for Biochemical Research*, 2nd Ed., 475, Oxford University Press, Oxford (1969)
18. *The Merk Index*, 9th Ed. (1976)
19. G. Bruening, R. Criddle, J. Preiss, and F. Rudert, *Biochemical Experiments*, 31, Wiley- Interscience, New York (1970)

POLY(CHLOROTRIFLUOROETHYLENE) SUBSTITUTION REACTIONS

Richard T. Taylor, J. A Shah, John W. Green and T. Kamolratanayothin

Chemistry Department
Miami University
Oxford, OH 45056

INTRODUCTION

Previous work by our group on polystyrene-based reagents showed that polystyrene could be mercurated and converted into polystyrene-bound phenyl seleninic acid, a versatile oxidizing agent[1]. Another polystyrene-based oxidant synthesized using the same mercuric intermediate was poly(iodoxystyrene)[2]. These polystyrene-based reagents were not hardy enough to allow repeated recycle of the polymeric reagents since benzylic protons, present in polystyrene, can be easily oxidized, causing the breakdown of the polymer backbone.

To overcome such a drawback, our group began a new program utilizing fluorocarbon polymers. One of the important properties of such polymers, contributed by the high fluorine content, is the resistance to reaction under a wide variety of conditions[3]. The use of these polymers would, therefore, allow us access to better oxidizing reagents. However, the desired inertness of fluoropolymers implies a synthetic challenge; the initial functionalization of fluoropolymers would be difficult.

Among the fluorine-containing polymers of commercial importance, our polymer of choice was PCTFE, the homopolymer of chlorotrifluoroethylene. The reason for choosing this polymer was the assumption that the chloride group would have sufficient reactivity to allow chemical modifications (Equation 1), but, in the most likely case that such modifications were incomplete, would be inert toward the ultimate reagents and substrates when the functionalized polymers were applied in their subsequent uses in carrying out organic reactions.

Early work by our group[4] and others[5-17] showed that the treatment of PCTFE with organolithium and organomagnesium reagents did result in loss of chloride and introduction of organic groups . However, substantial levels of side reactions took place. Among these were the formation of functionalized fluoroalkenes and oxygen incorporation. PCTFE was proposed to undergo a two-electron transfer mechanism which involved transmetallation of chloride by lithium (or magnesium) and then elimination of the metal fluoride, and readdition of the organic groups. Work by

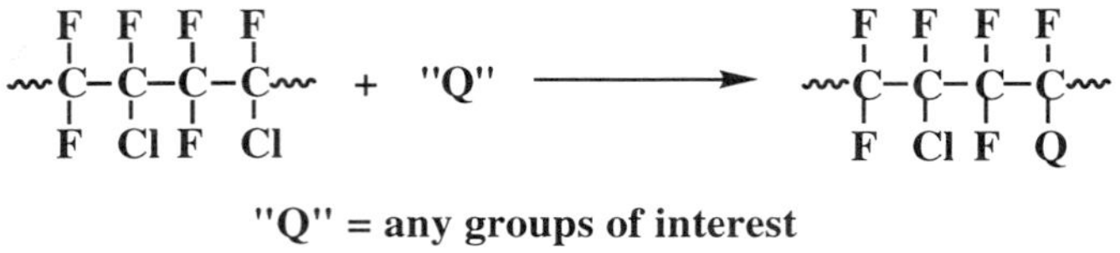

(1)

Cais[18] on tin hydride reduction of PCTFE to give poly(trifluoroethylene) (PTE) showed that the tin radical could selectively abstract the chlorine atom, but not the fluorine atom. The resulting PCTFE radical was trapped by hydride prior to any rearrangement. With these data in hand, our research group began to explore new ways to generate the PCTFE radical with subsequent trapping by other types of radical scavengers.

Our early work examined the reaction of PCTFE with sulfur, selenium and phosphorous nucleophiles[19] to achieve high levels of functionalization through a well-precedented (in the case of perfluoroalkyl iodides)[20-24] one electron transfer, radical anion chain process. While such a reaction demonstrated the feasibility of using one-electron processes for the functionalization of PCTFE, the carbon-sulfur linkage remained susceptable to oxidation.

For the purpose of arriving at polymeric reagents the formation of a carbon-carbon bond via radical means was desired. Trapping radicals in this way has been well known to modern organic chemists[25]. However, our problem in modifying PCTFE by this method was the heterogeneity of the polymer. In the solid state, PCTFE may be excluded from a solution-mediated reaction, and recovered unreacted.

Fortunately, our group was able to develop a strategy to overcome such a heterogeneity difficulty by using an allyltin species[26]. When a radical adds to allyltin, a tin radical would be produced. Therefore, the solution reaction would appear to keep regenerating the allyltin, and would not interupt the reaction between the polymer and the allyltin. In this fashion, treatment of PCTFE with allyltributyltin under free radical chain conditions gave allylated PCTFE with 10% degree of functionalization and with no detectable side reaction (Equation 2). Hence, our group proved that the above idea could be applied to the modification of PCTFE. However, such trapping agents are not easily devised. Consequently, a more general approach had to be found.

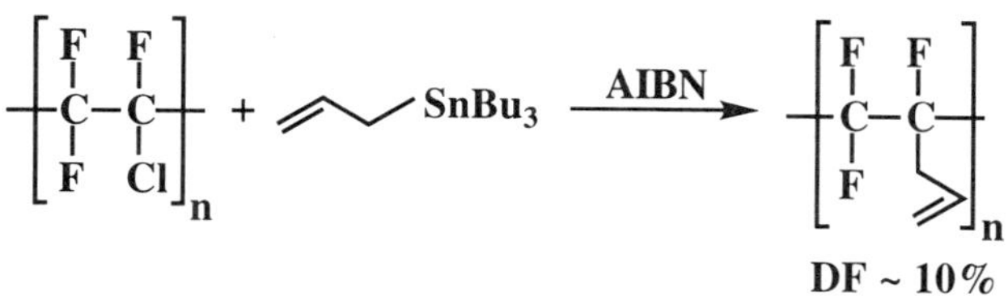

(2)

Some neutral metal carbonyls, particularly coordinatively unsaturated ones, are sufficiently nucleophilic to undergo reaction with organic halides. Sigma-alkyl species which undergo subsequent CO insertion are formed by such reactions[27]. Hence, carbonylations involving carbon monoxide insertion into metal-carbon sigma-bonds provide several useful synthetic approaches to complex organic carbonyl compounds[28]. In 1974, Lichstein[29] patented a process for the carbonylation of perfluoroalkyl iodides. The process involved reaction of a fluorocarbon iodide with carbon monoxide and an active-hydrogen containing compound in the presence of a catalytic amount of a metal carbonyl complex. The process is performed under a super-atmospheric pressure of carbon monoxide at a temperature between 150 and 200°C. The main product obtained is a carboxylic acid or a carboxylic acid derivative, depending upon the active-hydrogen containing compound employed. The patent describes the simple metal carbonyls, particularly $Cr(CO)_6$, $Mo(CO)_6$, $W(CO)_6$ and $Co_2(CO)_8$, as the preferred catalysts.

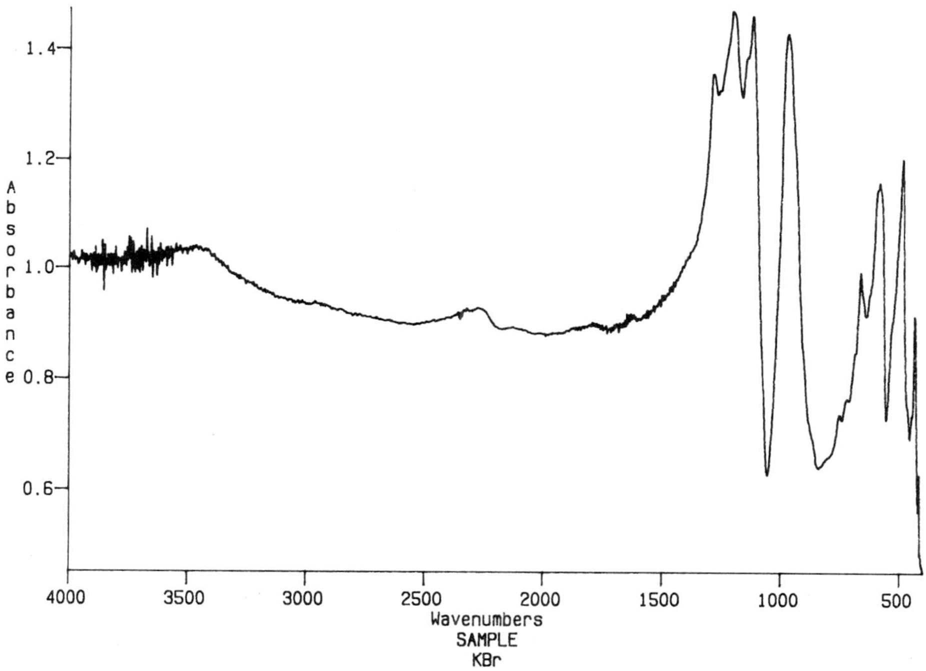

Figure 1. Absorption IR spectrum of Kel-F 6061.

In the following pages we describe our extensions in each of the above areas to afford sulfonic acid derivatives, carbon-carbon bond formation and carbonylation of PCTFE. The most direct indication of substitution is found through examination of the absorption IR spectrum. Changes in the spectrum of native PCTFE (see Figure 1) include loss of the carbon chlorine stretch and introduction of new stretching frequencies.

For the current study, the source of PCTFE was Kel-F 6061, a solid homopolymer of chlorotrifluoroethylene, produced by the 3M company. Preliminary studies with other forms of PCTFE indicate that its chemistry is by no means unique. We chose this particular polymer for its insolubility and other physical properties.

DISCUSSION

Reaction of PCTFE with Zn/SO_2

The production of industrially important perfluoroalkane sulfonic acids is generally accomplished by electrochemical fluorination. This method of preparation remains expensive and proceeds in good yields only for short hydrocarbon chains.[30] Recently however, Wakselman and Tordeux have described a chemical method for the preparation of trifluoromethane sulfonic acid.[31] The procedure involves reaction of a metal selected from zinc, cadmium, manganese, and aluminum with sulfur dioxide in DMF, followed by the introduction of trifluoromethyl bromide under slight pressure. The intermediate sulfinate is subsequently oxidized by hydrogen peroxide, and then hydrolyzed which leads to formation of the trifluoromethane sulfonic acid. Successful extension of the sulfination process to the modification of PCTFE should result in the formation of a sulfinated polymer which can ultimately be oxidized to give a sulfonic-acid modified polymer.

The sulfination procedure was adapted to the modification of PCTFE in the following manner. One equivalent of PCTFE and two equivalents of zinc were mixed in DMF and cooled to -23 °C in a dry ice/CCl_4 bath. An excess of sulfur dioxide gas was then added and the mixture was allowed to warm to room temperature, while being stirred, over a two-day period. The cloudy, yellow reaction mixture was quenched with saturated sodium hypochlorite solution, and the off-white product isolated by collection on a Büchner funnel. Analysis of the infrared spectrum for the derivative indicates that the desired sulfination occurred (Figure 2). Sulfinic acids show a weak band in the range 2800-2340 cm^{-1} associated with the OH stretching vibration of the $-SO_2H$ group. The position of the S=O stretching band is influenced by the electronegativity of the attached group. Electronegative substituents tend to raise the frequency, since they stabilize the form S=O rather than S^+-O^-. The normal range for the very strong S=O stretch extends from 1100-990 cm^{-1}. The S-O stretch, which varies in intensity from medium to strong, is located in the range of 870-810 cm^{-1}.[32] Inspection of the infrared absorption spectrum of the sulfinated PCTFE derivative reveals a weak, broad O-H stretch extending from 3000-2132 cm^{-1}. There is also a very strong S=O stretching band indicated at 1150 cm^{-1}. These bands are presumably shifted to higher than normal frequency due to the electronegativity of the perfluorocarbon backbone of the polymer. It appears that many of the modified sites are often isolated as the sulfonyl chloride. Sulfonyl chlorides absorb strongly at 1455-1405 cm^{-1} and also at 1225-1205 cm^{-1} due to the SO_2 asymmetric and symmetric stretching vibrations respectively.[32] The corresponding asymmetric SO_2 stretching band of the sulfinated PCTFE derivative is present as a strong absorption centered at 1454 cm^{-1}. Interpretation of the corresponding symmetric SO_2 stretch is complicated by the presence of the characteristic CF_2 stretching bands of the virgin PCTFE, but it appears to be present as a strong shoulder at 1230 cm^{-1}. The sulfonyl chloride is likely formed by attack of chloride anion on an intermediate in the formation of the sulfinate. The infrared spectrum also contains a weak band in the hydroxyl stretching region which extends from 3660-3356 cm^{-1}. This band is most likely due to water of hydration.[33]

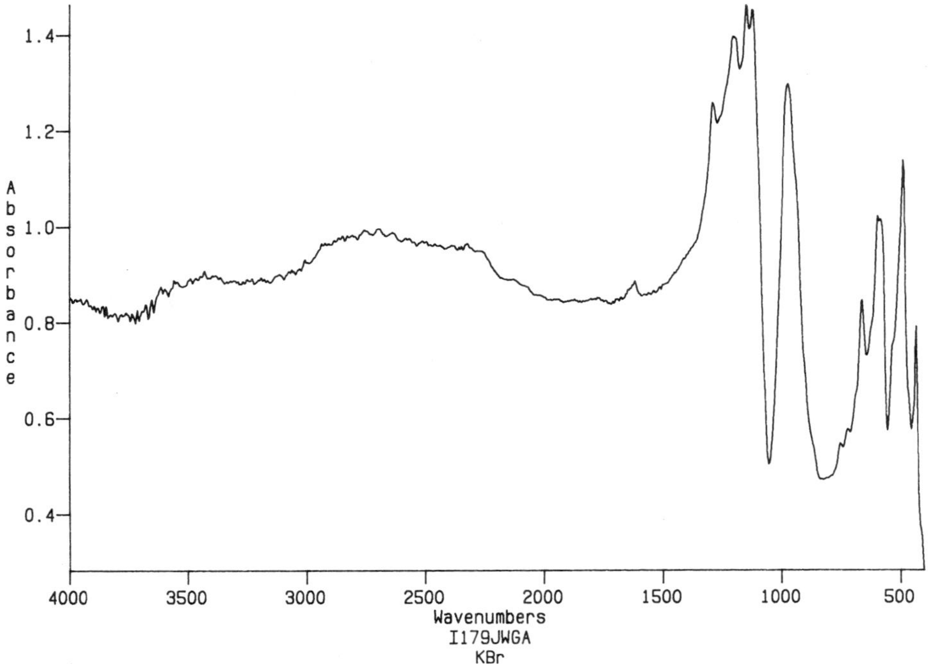

Figure 2. Absorption IR Spectrum of Sulfinylated PCTFE.

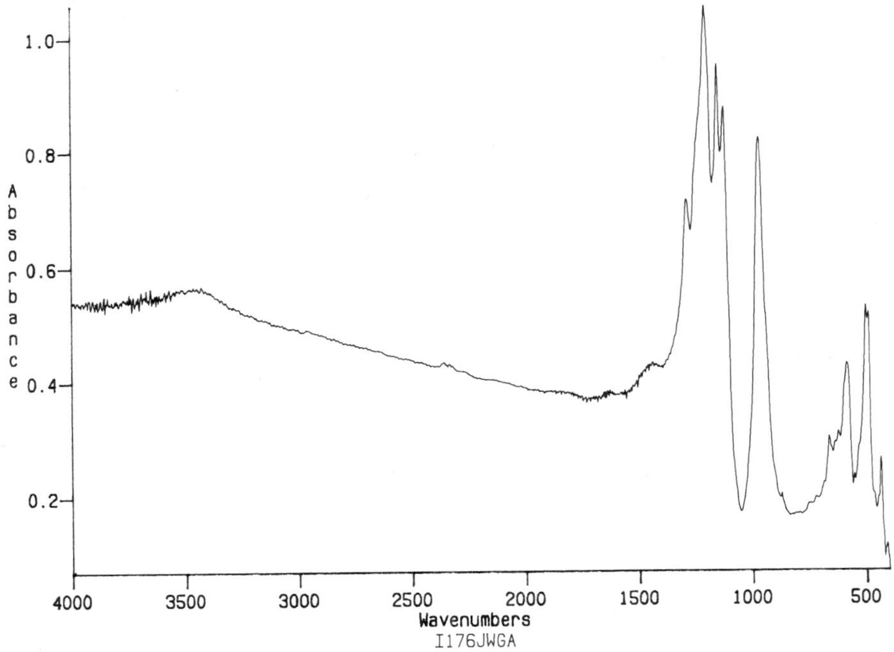

Figure 3. Absorption IR Spectrum of Sulfonylated PCTFE.

The appearance of all these bands in the infrared spectrum of the product is accompanied by a significant decrease in the characteristic C-Cl stretch at 970 cm^{-1}.

The sulfinated PCTFE derivative was converted to the sulfonated derivative by stirring the sulfinate in 30 wt % hydrogen peroxide/tetrahydrofuran for 2 days. The infrared spectrum (Figure 3) for the product polymer indicates formation of the sulfonated resin.

Small traces of water result in the ionization of sulfonic acids to the ionic $-SO_3^- H_3O^+$ form; therefore, observation of the covalent sulfonic acid by infrared analysis seems unlikely. The band due to the O-H stretching vibration of ionized sulfonic acids is of moderate intensity and very broad. It usually has several maxima in the range 2800-1650 cm^{-1}. The strong asymmetric and symmetric SO_3 stretching bands are located at 1230-1120 cm^{-1} and 1070-1025 cm^{-1} respectively. The infrared spectrum for the oxidized, sulfonated-PCTFE derivative does not show the O-H stretch characteristic of the covalently bound sulfonic acid, rather it indicates isolation of the sulfonic acid salt. The band due to the strong SO_3 asymmetric stretching vibration of sulfonic acid salts occurs at 1250-1140 cm^{-1}, while that associated with the symmetric stretching vibration is weaker, sharper and located at 1100-1030 cm^{-1}. For the product polymer, the asymmetric SO_3 stretch occurs at 1210 cm^{-1}, while that due to the symmetric SO_3 stretch is obscured by the CF_2 stretching vibrations characteristic of native PCTFE.

Reaction of PCTFE with Co(II) and Trapping Agents

As reported by Hu and Qiu[34] in 1992, bromo(pyridine)bis(dimethylglyoximato)cobalt(III) and zinc constituted a redox couple which could be used to promote perfluoroalkylation reactions of electron-deficient alkenes. This particular coordinated cobalt compound (bromo(pyridine)cobaloxime(III)) is a well-studied model compound of coenzyme vitamin B12. The starting materials used in the reactions were either perfluoroalkyl iodides or bromides, and the alkenes employed were ethyl acrylate, ethyl methacrylate, ethyl crotonate, acrylonitrile, and methyl vinyl ketone. These reactions were proposed to proceed via a perfluoroalkyl radical addition to a double bond. The low-valent cobalt (II) species, which was generated in situ by the Co(III)/Zn catalytic cycle was thought to initiate the formation of the perfluoroalkyl radical. Although, in our system, PCTFE is a chloride analog and is a polymer, a similar approach to the formation of the PCTFE radical with subsequent addition to a double bond might be expected to work as well. Therefore, we attempted to apply this method to the modification of PCTFE in our present work.[35]

Since bromo(pyridine)cobaloxime(III) was not commerically available and its synthesis was not convenient[36], we utilized chloro(pyridine)bis(dimethylglyoximato)-cobalt(III) (Equation 3) (also known as chloro(pyridine)cobaloxime (III)) instead. It has four cathodic waves in polarography when observed in acetonitrile. Its half wave potentials are located at -0.65, -1.45, -2.42, and -2.92 volts vs the Ag/$AgNO_3$ electrode, corresponding to the reduction of the cobalt from +3 to +2, +1, and 0, and the reduction of the ligand, respectively.

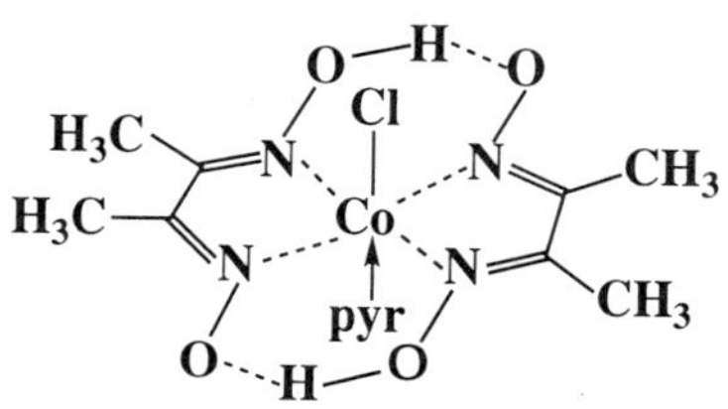

(3)

In order to find an optimum condition for producing a bimetal redox couple of chloro(pyridine)cobaloxime(III), several metals and solvents were investigated. The formation of the low-valent cobalt species was indicated by changing in color from brown to green when coupled with a reducing agent. Therefore, we used color to determine whether the redox couple of chloro(pyridine)cobaloxime (III) was obtained or not. It was found that the utilization of 1) zinc powder in either absolute ethanol, DMF, or THF; 2) lithium metal in either DMF or THF; and 3) magnesium (turnings) in DMF did not provide satisfactory results. The resulting mixtures failed to develop the desired green color. Fortunately, when magnesium (turnings) was used as a reducing agent, and THF as a solvent, chloro(pyridine)cobaloxime(III) could be reduced to afford the green mixture. Since the system was heterogeneous in nature, the time needed for the mixture to turn green varied, ranging from hours to days. However, the reduction could be facilitated to a certain extent by heating the mixture to reflux.

One might think that changing the reducing agent from zinc to magnesium might lead to an undesired consequence since it has been well known that alkyl halides can react with magnesium to produce Grignard reagents. However, in our system, we had no concern about that because both magnesium and PCTFE were in the solid state. Therefore, they would not react with each other to give a Grignard reagent of PCTFE, which could undergo the two-electron transfer process, as mentioned earlier. As a result, we were confident that the two-electron transfer process was avoidable in our particular case.

Once we were able to prepare the green suspension of chloro(pyridine)-cobaloxime(III) and magnesium, the addition of PCTFE to a series of alkenes under this reductive condition could be examined. However, prior to investigation of these addition reactions, we conducted two different experiments, as blanks, in the absence of participating alkenes. One experiment involved a stoichiometric amount of chloro(pyridine)cobaloxime(III) (1.0 equivalent versus Cl), and the other employed the catalytic quantity (0.1 equivalent). We found that, in both cases, PCTFE reacted with the mixture of chloro(pyridine)cobaloxime(III), magnesium and THF to give PCTFE with partial loss of chloride and incorporation of oxygen and organic moieties.

In the IR spectrum of the product from the blank employing one equivalent of chloro(pyridine)cobaloxime(III), an oxygen moiety was observed as broad bands in two regions, 3200 - 3700 and 1600 - 1800 cm^{-1}. The presence of the hydrocarbon chain was apparent at 2990, 2891, 1398, and 1355 cm^{-1}. In the case of the blank, using the catalytic amount of chloro(pyridine)cobaloxime(III), the IR spectrum of the product contained a broad band in 3200 - 3700 cm^{-1} region and a weak band at 1632 cm^{-1}, indicating the presence of the oxygen component, and weak bands at 3002 and 2964 cm^{-1} with a medium band at 1384 and 1352 cm^{-1} for the organic portion.

The extent of the above so-called side reaction was determined by considering the decrease in the absorbance of the chloride band. We found that with the stoichiometric amount of chloro-(pyridine)cobaloxime(III), the side reaction was much more prominent than with the catalytic amount. Nevertheless, no appreciable unsaturation (the absence of $CF{=}CF_2$ around 1790 cm^{-1})[39] was indicated in the IR spectra in both cases, implying that the undesired two-electron transfer process was not involved.

From the above discovery, two conclusions might be drawn. One was that the reactive form of PCTFE, perhaps as a PCTFE radical, could be generated by using the mixture of chloro(pyridine)-cobaloxime(III) and magnesium. This assumption led to the second conclusion which was that the incorporation of oxygen and organic moieties in the resulting product might arise from the addition of the PCTFE radical to THF, and possibly the THF hydrogen atom abstraction of the PCTFE radical. When the stoichiometric amount of chloro(pyridine)cobaloxime(III) was used, more PCTFE radicals were formed, compared with those produced in the catalytic fashion. As a result, a higher degree of the side reaction was seen in the former case.

Based on the above conclusions, three consequences might be expected when the bimetal redox couple was applied to the modification of PCTFE. By employing the alkenes which were more reactive toward the addition to PCTFE than THF was, we might be able to prevent the side reaction, and obtain the desired product. Contradictorily, if the alkenes used were not reactive enough, the side reaction would predominate, and as a result, we might either obtain the addition product in conjunction with the product from the side reaction, or the product from the side reaction solely. In the last possible case, the mixture of chloro(pyridine)cobaloxime(III) and magnesium might not only initiate the formation of the reactive form of PCTFE, but also initiate the polymerization of the alkenes used. This would result in either the grafting of the alkene oligomer onto PCTFE or, in the extreme case, unreacted PCTFE. Therefore, the reactivity of the alkene would determine the fate of the reaction.

Our first choice of alkene was allyltributyltin which had been previously studied by our group (see above). By employing the chloro(pyridine)cobaloxime(III)/magnesium redox couple in which chloro(pyridine)cobaloxime(III) was used in the catalytic amount, we were now able to synthesize the same allylated PCTFE (Eq. 4) with a higher degree of functionalization of approximately 50%.

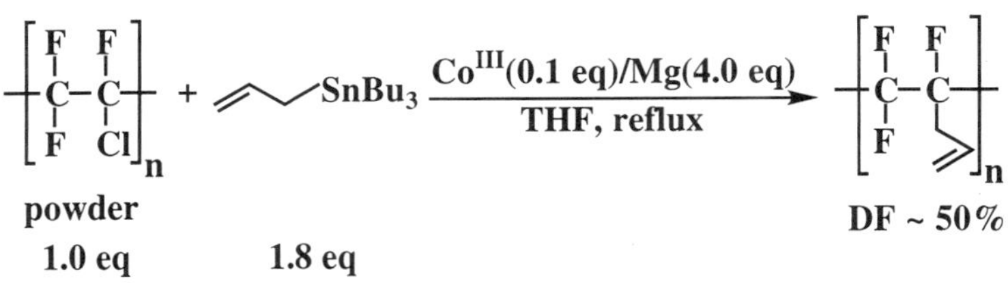

(4)

According to the elemental analysis of allylated PCTFE (C, 34.48%; H, 2.32%; Cl, 13.28%; F, 46.97%), the combined percentage for carbon, hydrogen, chlorine, and fluorine elements was 97.05. In addition, its IR spectrum (Figure 4) clearly showed the presence of the allyl moiety at 3087 (=CH stretching), 2982, 2886 (CH stretching) and 1643 (C=C stretching) cm^{-1} with concomitant loss of the chloride band at 972 cm^{-1}. Therefore, based on both IR and elemental analyses, we were confident that the addition of PCTFE across the double bond of allyltributyltin was clean, and the subsequent loss of tributyltin species, perhaps as a radical, resulted in allylated PCTFE.

It should be noted that the similar addition did not succeed when a stoichiometric amount of chloro(pyridine)cobaloxime(III) was utilized. We found that the reaction of PCTFE did occur, but with a significant level of the side reaction (broad bands in 3100 -3700 and 1500 - 1800 cm^{-1} regions in the IR spectrum. The rationale of the result was that a large number of PCTFE radicals were rapidly produced at one time. Some of these radicals were, therefore, available to react with THF concurrently with the desired addition to allyltributyltin.

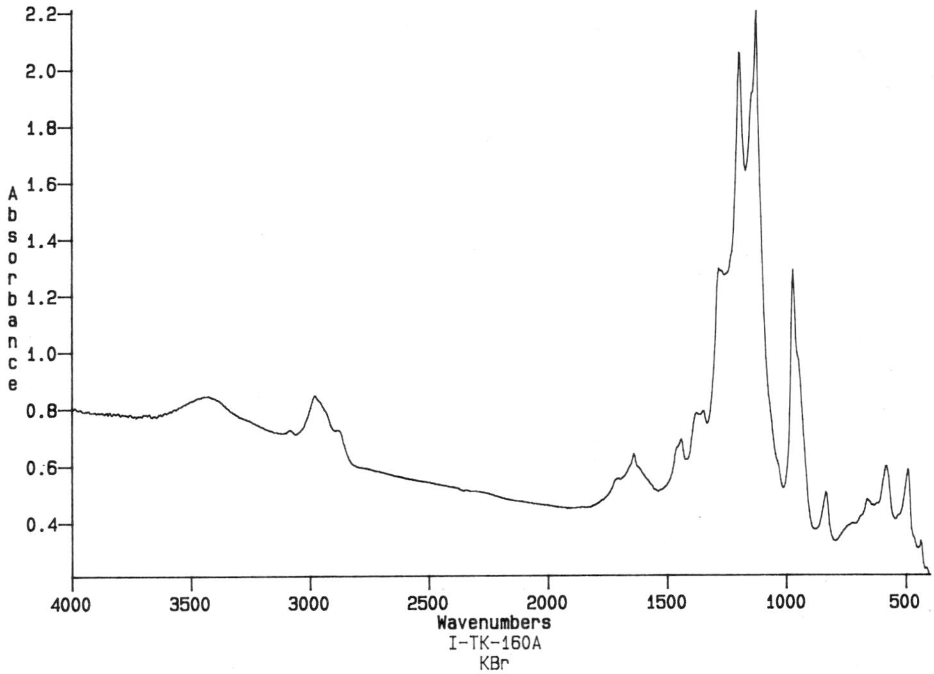

Figure 4. Absorption IR Spectrum of Allylation of PCTFE.

After the successful addition of PCTFE to allyltributyltin, the next alkene we examined was styrene. By treating PCTFE with styrene in the presence of magnesium and a catalytic amount of chloro(pyridine)cobaloxime(III), phenethylated PCTFE was obtained as depicted in Equation 5.

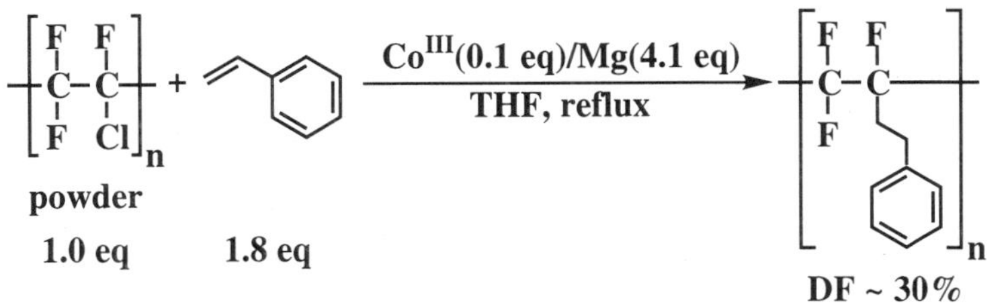

(5)

In the IR spectrum of the resulting polymer (Figure 5), bands at 3087, 3064, and 3036 cm^{-1} indicated the aromatic C-H, and bands at 1602, 1496, 746, and 700 cm^{-1} confirmed the presence of the aromatic ring with monosubstituted pattern. The elemental analysis of the polymer (C, 38.67%; H, 1.87%; Cl, 17.79%) suggested that one styrene rather than its grafted oligomer was added per chlorine removed, giving a degree of functionalization of approximately 30%.

Once again, it should be noted that the reductive condition in which the stoichiometric amount of chloro(pyridine)cobaloxime(III) was used, not only aided the addition of PCTFE to styrene, but also promoted the side reaction. In the IR spectrum of the resulting polymer, the reduction in the absorbance of the chloride band at 972 cm^{-1} seemed to be too large, and the incorporation of oxygen in 3300 - 3600 cm^{-1} region was also evident. The situation similar to the case of allyltributyltin might be applied in that THF possibly also took part in the reaction.

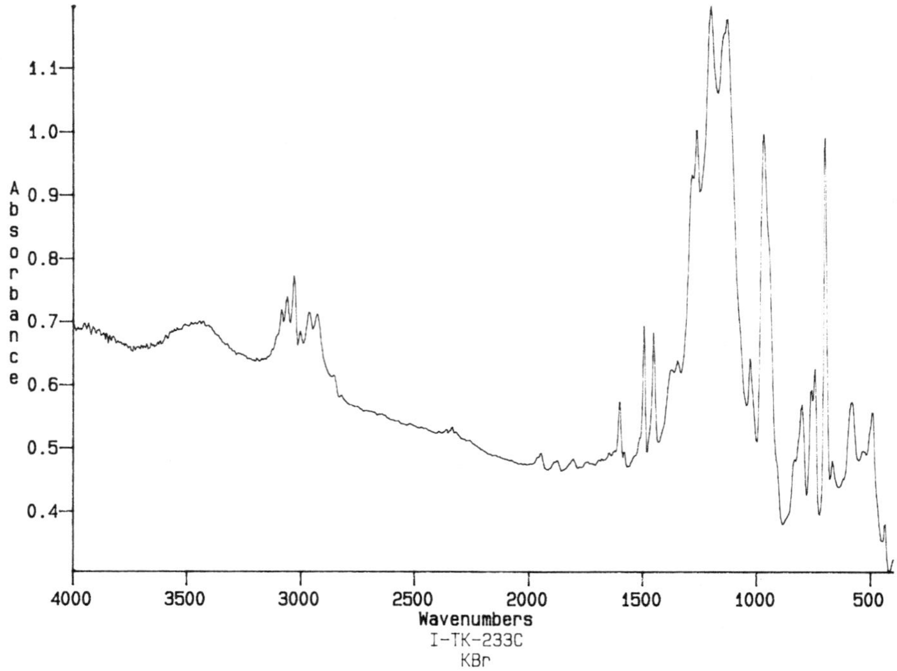

Figure 5. Absorption IR spectrum of phenethylated PCTFE.

The fact that allyltributyltin and styrene which are electroneutral were successfully added to PCTFE led us to the question "Can electron-rich or electron-poor trapping agents be used?". The electron-rich agents that we examined were ethyl ethynyl ether and ethyl vinyl ether. On the side of the electron-poor alkenes, eight were investigated: ethyl acrylate, methyl methacrylate, methyl vinyl ketone, acrylonitrile, methacrylonitrile, vinyl bromide, chloromethyl styrene, and 4-vinylpyridine. The details of each reaction are summarized.

In the case of the electron rich reagents, no addition products were observed, either under catalytic or stoichiometric conditions. In all cases, no evidence of incorporation was discerned from IR analysis. In addition to unidentifiable materials (in the ethyl ethynyl ether case), substantial side reaction could be recognized.

In the case of the electron poor alkenes, results were more varied. Under all conditions examined, reactions with methyl vinyl ketone, acrylonitrile, methacrylonitrile and 4-vinyl pyridine afforded products with IR spectra equivalent to those obtained without the addition of the alkene (side reaction). In the cases of vinyl bromide and chloromethyl styrene, unreacted PCTFE was recovered unchanged. It is speculated that electron transfer to the alkene proceeded in each case. While the product of vinyl bromide reduction was not observed, perhaps because of volatility, one could isolate poly(chloromethylstyrene) in the latter case.

We discovered that PCTFE could be added to ethyl acrylate when a stoichiometric amount of chloro(pyridine)cobaloxime(III) was used. In accordance with the IR spectrum of the resulting polymer, the carbonyl band was apparent at 1735 cm^{-1} which corresponded to the C=O group in a saturated ester. Unfortunately, the decrease in the absorbance of the chloride band at 971 cm^{-1} seemed to be too large, compared with the increase in the absorbance of the carbonyl band, implying that the side reaction also occurred.

Since the stoichiometric condition did provide the addition of PCTFE to ethyl acrylate but along with the side reaction, we expected that the chemospecificity of the addition might be improved by changing the condition to the catalytic one. This assumption was based on the result observed in the blanks -- the less the amount of chloro(pyridine)cobaloxime(III) used, the lower the degree of the side reaction.

It was shown that by employing the catalytic quantity of chloro(pyridine)cobaloxime(III) (Equation 6), the chemo-specificity of the addition was enhanced. In accordance with the IR spectrum of the functionalized PCTFE (Figure 6), the carbonyl band at 1736 cm^{-1} was much larger while the reduction in the absorbance of the chloride band at 972 cm^{-1} was much smaller, compared with those observed in the stoichiometric condition. However, the side reaction was not completely suppressed as the bands due to the side reaction were still present.

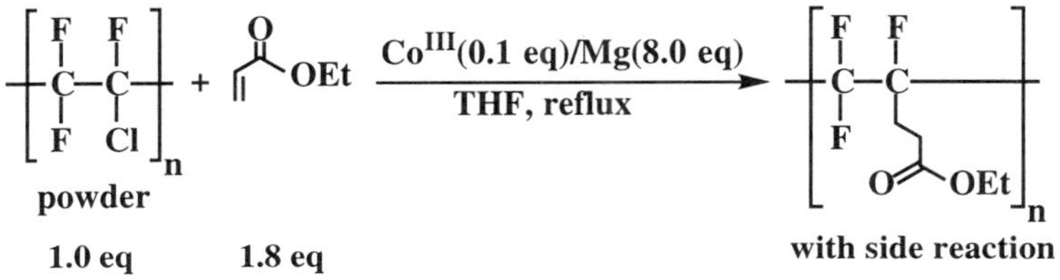

(6)

In order to prove that the presence of the carbonyl band in the IR spectrum was not due to polyethyl acrylate, we did the blank of ethyl acrylate, chloro(pyridine)cobaloxime(III) (in a catalytic amount), magnesium, and THF. No polymer was obtained.

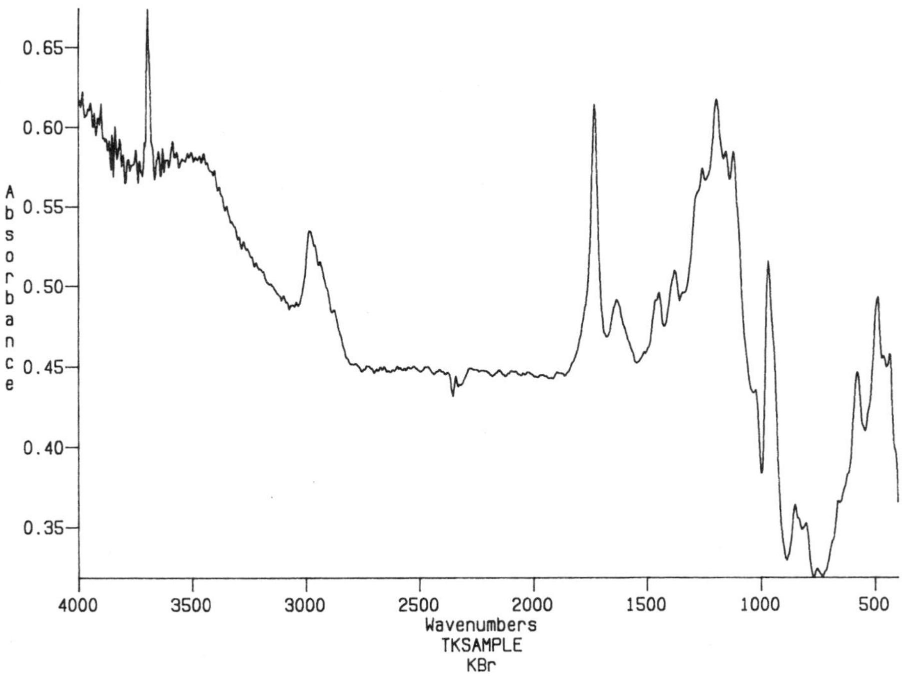

Figure 6. Absortion IR spectrum of addition of PCTFE to Ethyl Acrylate.

PCTFE was found to add to methyl methacrylate via the use of a stoichiometric amount of chloro(pyridine)cobaloxime(III). In the IR spectrum of the resulting product, the carbonyl group of a saturated ester was shown as a strong band at 1735 cm^{-1}. However, the loss in the chloride band at 975 cm^{-1} was tremendous, and strong, broad bands due to the side reaction in 3100 - 3700 cm^{-1} region and at 1629 cm^{-1} were observed as well.

Similar to the case of ethyl acrylate, the chemospecificity of the addition of PCTFE to methyl methacrylate was improved by using a catalytic amount of chloro(pyridine)cobaloxime(III). The IR spectrum of the resulting polymer showed the carbonyl band at 1731 cm^{-1}. Although the bands due to the side reaction were also present, the decrease in the absorbance of the chloride band at 974 cm^{-1} became less obvious, inferring that the extent of the side reaction was lowered.

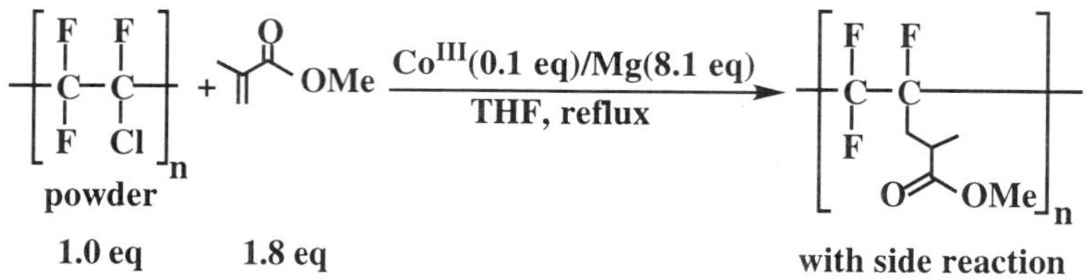

(7)

In summary, inspection of these results indicate that the range of suitable alkene trapping agents is rather narrow and is restricted to relatively electroneutral and weakly electron poor species. The electron poor nature of the fluoropolymer radical generated from reduction implies slow reaction with electron poor trapping agents, giving rise to increasing

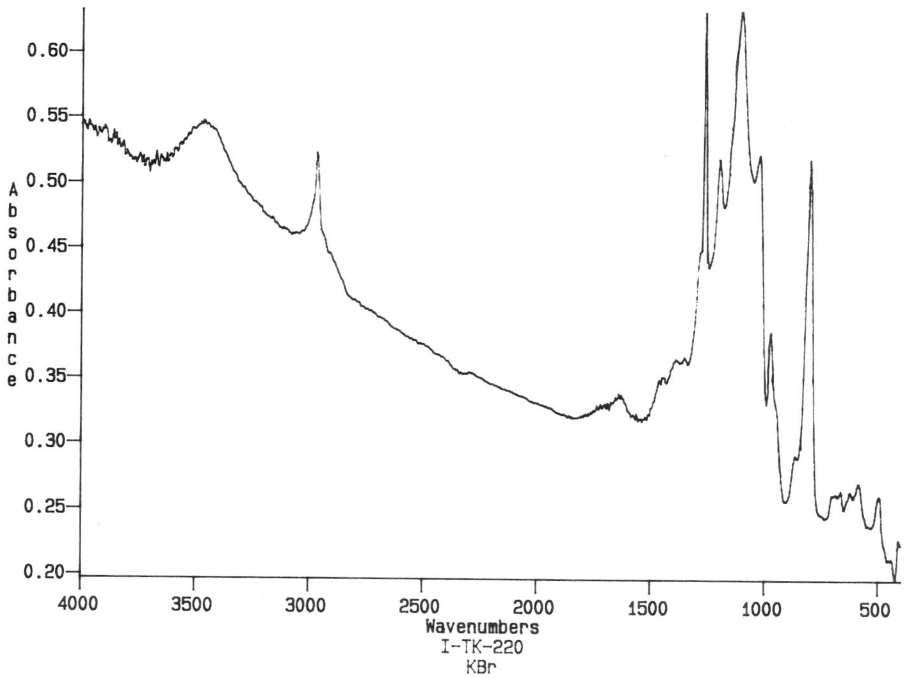

Figure 7. Absorption IR Spectrum of Hydroboration-Oxidation of Allylated PCTFE.

amounts of side reaction. Addition of the radical to an electron rich alkene may well be reversible, giving rise to the side reactions as well.

While the range of effective trapping agents was rather limited, those most successful were amenable to further reaction. By taking advantage of the alkene functionality present in allylated PCTFE, the polymer could be subjected to further functionalization. It was found that the hydroboration-oxidation of the double bond in allylated PCTFE afforded PCTFE-1-propanol (Equation 8). According to the IR spectrum of the resulting polymer (Figure 7), the O-H stretching appeared as a broad band with its maximum at 3445 cm^{-1}, and the C-O stretching was observed at 1026 cm^{-1}.

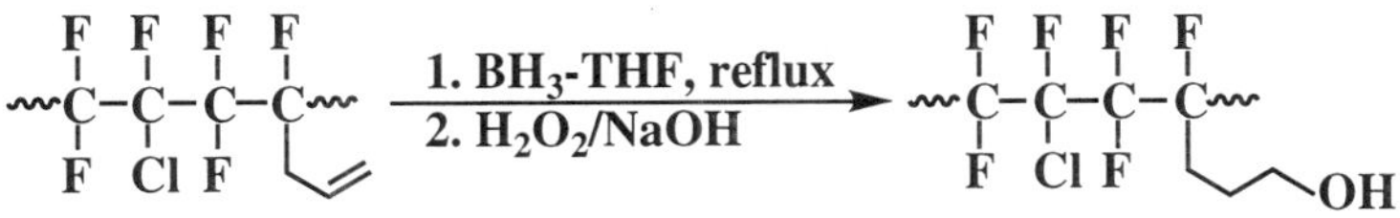

(8)

With pyridinium dichromate[37], the oxidation reaction of PCTFE-1-propanol produced PCTFE-propanoic acid as illustrated in Equation 9. Its IR spectrum showed the presence of the carboxylic acid moiety at 3439 (broad, O-H stretching) and 1656 (broad, perhaps asymmetric stretching of two carbonyl groups of carboxylic acid salt) cm^{-1}. Furthermore, the disappearance of C-O and =C-H bands was clearly observed in the IR spectrum.

(9)

By treating styrylated PCTFE with concentrated nitric and sulfuric acids, the corresponding nitrated polymer was obtained (Equation 10). In accordance with the IR spectrum of the resulting polymer (Figure 8), bands at 1529 and 1350 cm^{-1} indicated the nitro group and a band at 856 cm^{-1} confirmed the presence of the nitroaromatic compound. However, the extent of nitration was not determined.

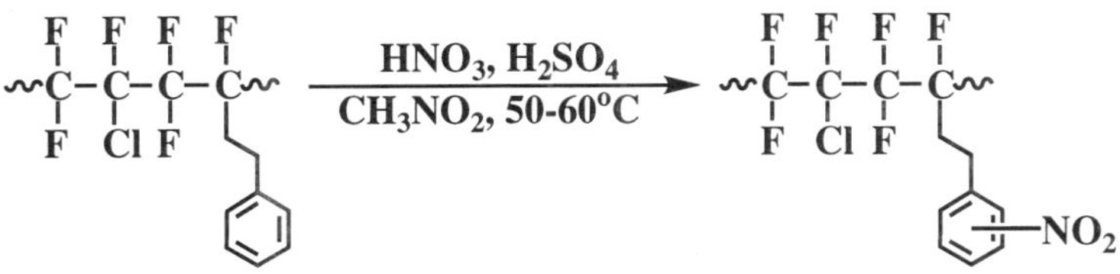

(10)

Reaction of PCTFE with Metal Carbonyls

Reaction of PCTFE with a stoichiometric amount of chromium hexacarbonyl in DMF at 95°C for 5 days under a nitrogen atmosphere, followed by hydrolysis results in the formation of a brown-black polymer. Analysis of the infrared data indicates that carbonylation does indeed occur (Equation 11). The infrared absorption spectrum shows a large decrease in the C-Cl stretch at 970 cm^{-1} with a concomitant appearance of a very strong band in the carbonyl stretching region centered at 1680 cm^{-1}. There is also a broad band centered at 3490 cm^{-1} in the hydroxyl stretching region and two bands of moderate intensity

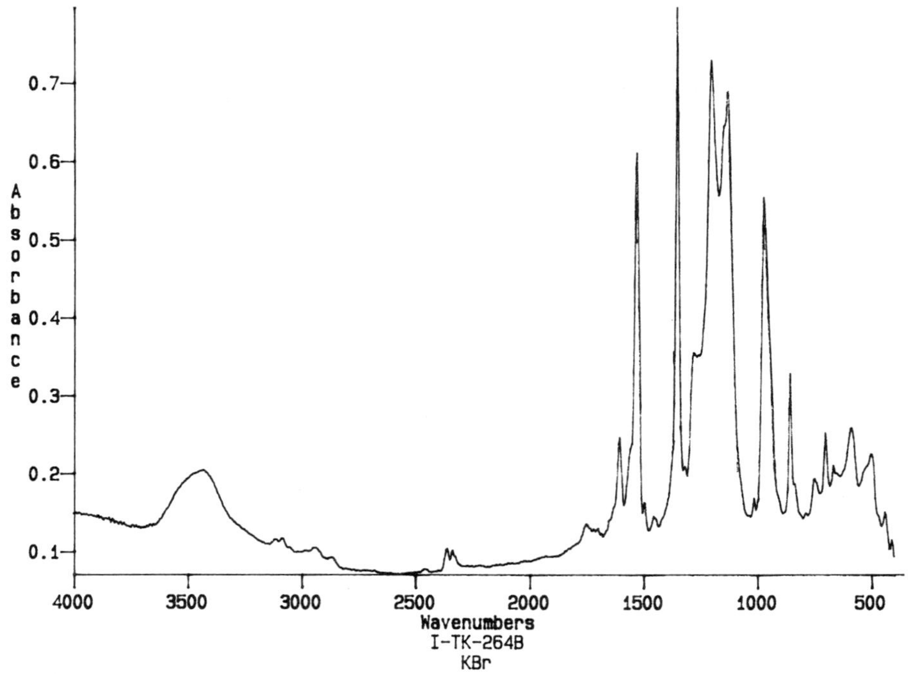

Figure 8. Absorption IR Spectrum from Nitration of PCTFE Addition to Styrene.

centered at 1400 and 1355 cm^{-1}, respectively. The characteristic bands due to the symmetric and antisymmetric CF_2 stretching are also present between 1300 and 1100 cm^{-1} (Figure 9). Closer examination of the infrared data indicates that the expected carboxylic-acid modified PCTFE has been isolated in the carboxylate form. The carboxylate anion generally gives rise to two bands: a strong asymmetrical stretching band near 1650-1550 cm^{-1} and a weaker, symmetrical stretching band near 1400 cm^{-1}. Substitution in the alpha position with an electronegative group such as the fluorines present in PCTFE, leads to a slight increase in the C=O absorption frequency of 10-20 cm^{-1}.[38]

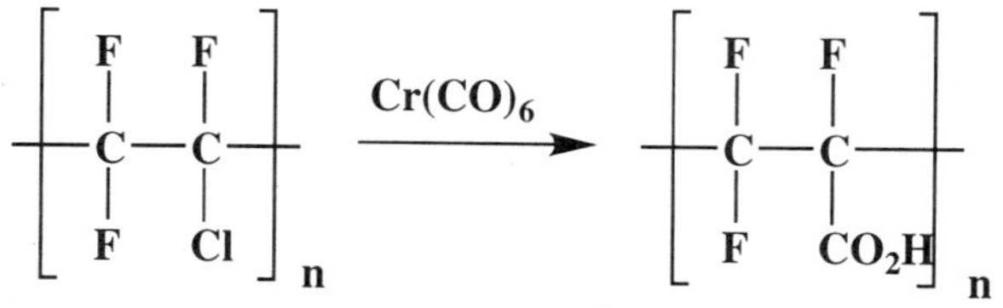

(11)

Further support for the presence of the carboxylate group is found on comparison of the infrared data to that collected for low-molecular-weight perfluorocarboxylate salts. Huang, Haas, and Lieb[39] prepared sodium perfluorocarboxylates by stirring perfluoroalkyl iodides or bromides with sodium formaldehyde sulfoxylate (Rongalite) in the presence of sodium bicarbonate in aqueous DMF. Infrared data for several of the low-molecular-weight analogs are in close agreement and the presence of a carboxylate anion is clearly indicated in the PCTFE derivative. The broad band centered at 3490 cm^{-1} in the hydroxyl group stretching region is presumably due to water of hydration. The presence of hydrated ion clusters in organic ion exchangers is well studied.[33]

The carbonylation of PCTFE with chromium hexacarbonyl appears to have a definite temperature dependence. The optimum temperature range for clean and efficient carbonylation of PCTFE is from 90 to 100°C. At temperatures below 90°C, only a low degree of functionalization occurs. For example, reaction of a PCTFE with a stoichiometric amount of $Cr(CO)_6$ in DMF at 80°C for 7 days, followed by subsequent hydrolysis led to the isolation of a very light brown powder. Analysis of the infrared spectrum shows that essentially unreacted PCTFE is recovered. Reaction of PCTFE at temperatures higher than 100°C results in reduction of the polymer backbone, as indicated by the appearance of bands at 2975, 2940, and 2875 cm^{-1} in the C–H stretching region of the infrared spectrum. As long as the reaction temperature is maintained in the optimum range, formation of significant amounts of the reduced by-product appears to be avoided.

By employing active-hydrogen containing compounds other than water in the carbonylation procedure, Lichstein was able to prepare a wide variety of carboxylic acid derivatives[29]. The main product received is a fluorinated ester, anhydride, thioester or amide, depending upon the active-hydrogen containing compound employed. Successful extension of this chemistry to PCTFE should result in a wide range of functionalized derivatives which may prove useful as reactive polymers or as solid phases in chromatography. Hence, a systematic investigation of the reaction of PCTFE with $Cr(CO)_6$ and various active-hydrogen containing compounds was undertaken in our laboratories. The results are discussed in the following paragraphs. The choice of active-hydrogen compounds

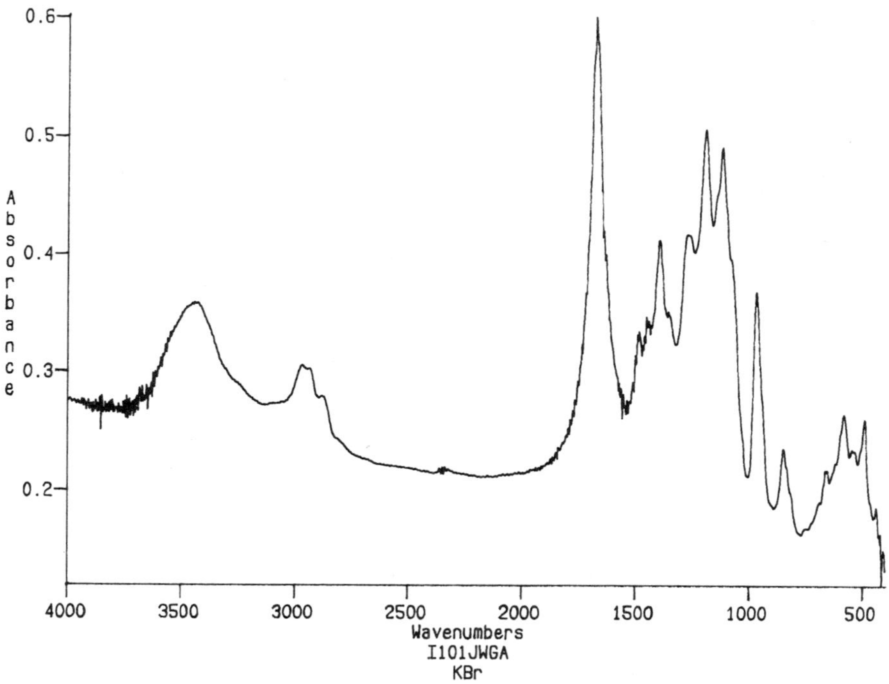

Figure 9. Carbonylation of PCTFE with $Cr(CO)_6$ and Water.

which may be employed is somewhat limited by the optimum reaction conditions. Since the reaction does not proceed to any appreciable degree unless a temperature above 90°C is maintained, active-hydrogen compounds with boiling points below 90°C cannot be used if efficient functionalization is to be achieved.

The reaction of PCTFE with $Cr(CO)_6$ and active-hydrogen-containing amines results in the formation of amides (Equation 12). The reaction proceeds with either primary or secondary amines, and the substituents may be of either an aliphatic or aromatic nature. When PCTFE is treated with an appropriate amine and $Cr(CO)_6$ in DMF at 90°C for 4 days, a derivative is produced for which infrared spectral data indicate the presence of the corresponding amide moiety. Infrared spectral data for the secondary amide formed by reaction with cyclohexylamine indicate that the desired reaction occurs (Figure 10). A moderate band due to the N-H stretch is centered at 3390 cm^{-1}. Aliphatic C-H stretches of the cyclohexyl group are centered at 2931 and 2885 cm^{-1}. Overlap of the C=O stretch, (amide I band), and N-H bend, (amide II band), is seen in a strong, broad band ranging from 1700-1620 cm^{-1}. A strong band due to the C-N stretch is observed at 1454 cm^{-1}. The appearance of all these bands is accompanied by a significant decrease in the C-Cl stretch of PCTFE at 970 cm^{-1}.

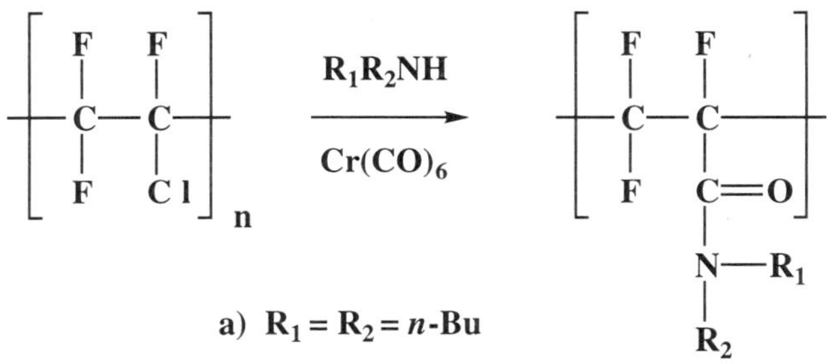

(12)

a) $R_1 = R_2 = n$-Bu

b) R_1 = H; R_2 = Cyclohexyl

c) R_1 = H; R_2 = Benzyl

Elemental analysis for this amide derivative, case b, indicates that side reactions occur. Elemental analysis shows that C (39.06%), H (3.66%), Cl (11.37%), F (24.45%) and N (6.12%) comprise 84.66% of the mass of the polymer. It is assumed that oxygen comprises the remaining 15.34% of the mass of the polymer. Assuming the derivative is an idealized copolymer, a degree of functionalization of 84% is calculated based on the %N reported in the elemental analysis. However, a degree of functionalization of 50% is calculated based on the %H reported in the elemental analysis. Thus, competing carbonylation and/or elimination clearly skew the results. In any case, the polymer is clearly functionalized to a significant extent.

Formation of an amide is also indicated in the reaction of PCTFE with $Cr(CO)_6$ and the primary amine, benzylamine. The infrared absorption spectrum shows an N-H stretch centered at 3400 cm^{-1}, aromatic C-H stretches at 3063 and 3030 cm^{-1}, aliphatic C-H stretches at 2933 and 2876 cm^{-1}, a broad amide I/amide II band ranging from 1680-1580 cm^{-1}, and a C-N stretch at 1454 cm^{-1}. The C-Cl stretch at 970 cm^{-1} also shows a significant decrease in

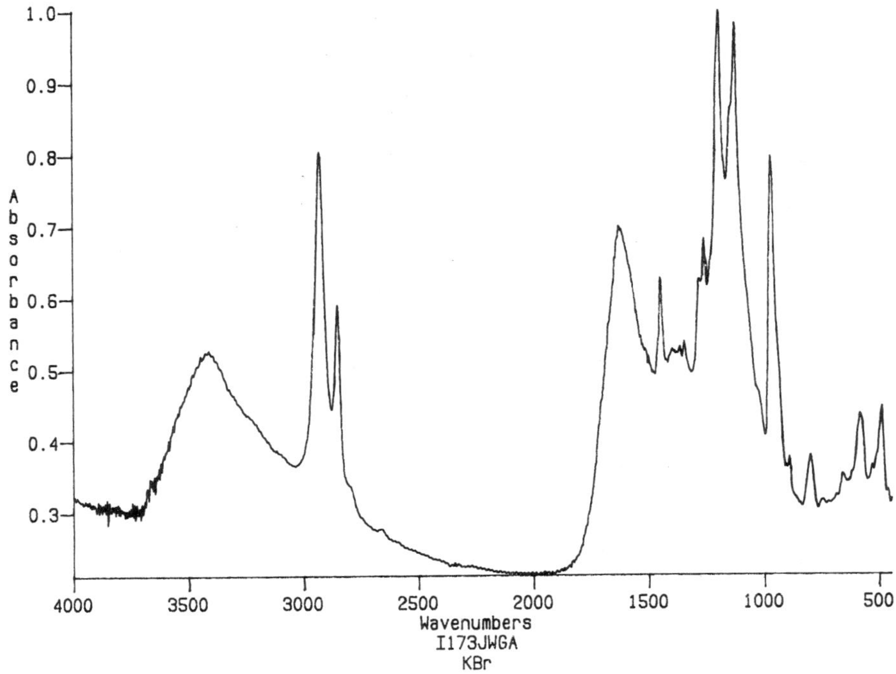

Figure 10. Carbonylation of PCTFE with $Cr(CO)_6$ and Cyclohexylamine.

intensity. Similarly, when PCTFE is reacted with $Cr(CO)_6$ and the secondary amine dibutylamine, the resulting product appears to be the corresponding amide. The presence of the butyl groups is indicated by bands in the IR spectrum at 2964, 2936, and 2877 cm^{-1}. The carbonyl stretching absorption occurs in the range from 1680-1600 cm^{-1}, and the C-N stretch occurs at 1464 cm^{-1}. The C-Cl stretch again shows a significant decrease in intensity. Interestingly, a broad band of moderate intensity centered at 3440 cm^{-1} is also observed in the infrared spectrum. This indicates that the sample has water of hydration, which implies that water acts as the active-hydrogen compound to form the carboxylate derivative in a competing reaction. Formation of the carboxylated derivative is likely a competing process in all cases and its formation may be accounted for based on steric reasons. A small water molecule can clearly more readily gain access to the PCTFE/chromium complex than a bulky primary or secondary amine, if only steric effects are in operation. The water can then react with the complex to form the carboxylated derivative. Thus, a variety of amide derivatives may be prepared by this methodology, but if water is not carefully excluded from the reaction, competing hydrolysis can be significant.

Reaction of PCTFE with $Cr(CO)_6$ and an alcohol in attempts to prepare the corresponding ester met with more limited success. When PCTFE is stirred with methanol and $Cr(CO)_6$ in DMF at 60°C for 5 days, no appreciable reaction occurs. Infrared analysis of the product shows that essentially unreacted starting material is recovered. When the same reaction is repeated at 90°C, the sole product appears to be the carboxylated PCTFE derivative. Whatever amount of methanol may have remained dissolved in the DMF under these conditions, does not appear to result in the formation of significant amounts of the methyl ester.

When higher boiling alcohols were employed in a similar reaction, essentially no reaction occurred. Reaction of PCTFE with $Cr(CO)_6$ and either 2-naphthol, isobutyl alcohol, or *tert*-butyl alcohol results in the formation of a light brown polymer. Infrared analysis in each case reveals the product to be essentially unreacted starting material. Because of the steric bulk of the alcohols employed, it seems likely that reduced accessibility to the reacting sites may contribute significantly to the lack of success of the reaction. The preparation of thioesters was also disappointing. Reaction of PCTFE with $Cr(CO)_6$ and thiophenol resulted in the formation of a light brown polymer. In the infrared spectrum, weak bands at 3050 and 2937 cm^{-1} indicate the presence of aromatic C-H. A carbonyl band is located at 1686 cm^{-1}, but contamination in the form of metal carbonyls is indicated by bands at 1982, 2273, 2338, and 2362 cm^{-1}. There is only a small decrease in the intensity of the C-Cl stretch at 975 cm^{-1}, so a large degree of functionalization is not indicated.

Attempts to prepare acid anhydride derivatives by our methodology were unsuccessful. Various carboxylic acids were reacted with PCTFE in the presence of $Cr(CO)_6$ in efforts to prepare anhydrides, but no appreciable reaction occurred in any case. The major side reaction in each case appears to have been formation of the carboxylated PCTFE derivative.

CONCLUSION

The strategy of using one electron processes to effect the substitution of PCTFE can result, under the right circumstances, in a clean substitution with high degree of functionalization. The major difficulties lie in competing reactions with solvent and in two electron reduction leading to elimination. Future efforts in our research group will include examination of the material properties (particularly in ion exchange), further new substitution reactions and reaction of already substituted materials.

ACKNOWLEDGEMENTS

We gratefully acknoweledge the support provided by the Miami University Reseach Advisory Council in the form of a Shoupp grant, as well as the 3M Corporation, for their gift of a generous supply of Kel-F 6061.

REFERENCES

1. Taylor, R. T.; Flood, L. A. *J. Org. Chem* **1983**, *48*, 5160.
2. Stevenson, T. A.; Taylor, R. T. *Reactive Polymers* **1988**, *8*, 7.
3. Wall, L. A. *Fluoropolymers*; Wiley-Interscience: New York, 1971;Chapter 16, pp. 507-543.

4. Danielson, N. D.; Taylor, R. T.; Huth, J. A.; Siergiej, R. W.; Galloway, J. G.; Paperman, J. B. *Ind. Eng. Chem., Prod. Res. Dev.*, **1983**, *22*, 303.
5. Huth, J. A.; Danielson, N. D. *Anal. Chem.* **1982**, *54*, 930.
6. Siergiej, R. W.; Danielson, N. D. *Anal. Chem.* **1983**, *55*, 17.
7. Kruempelman, M.; Danielson, N. D. *Anal. Chem.* **1985**, *57*, 340.
8. Dias, A. J.; McCarthy, T. J. *Polym. Prepr.* **1985**, *26*, 70.
9. McCarthy, T. J.; Dias, A. J. *Macromolecules* **1985**, *18*, 1826.
10. Dias, A. J.; McCarthy, T. J. *Polym. Prepr.* **1986**, *27*, 68.
11. Lee, K. -W.; McCarthy, T. J. *Macromolecules* **1987**, *20*, 1437.
12. Dias, A. J.; McCarthy, T. J. *Macromolecules* **1987**, *20*, 2068.
13. Bee, T. G.; McCarthy, T. J. *Polym. Mater. Sci. Eng.* **1990**, *63*, 94.
14. Shoichet, M. S.; McCarthy, T. J. *Polym. Prepr.* **1990**, *31*, 418.
15. Shoichet, M. S.; McCarthy, T. J. *Macromolecules* **1991**, *24*, 982.
16. Bee, T. G.; McCarthy, T. J. *Macromolecules* **1992**, *25*, 2093.
17. Beaver, L. G.; Danielson, N. D. *Polymer Bulletin* **1993**, *30*, 47.
18. Cais, R. E.; Kometani, J. M. *Macromolecules* **1984**, *17*, 1932.
19. Pelter, M. W.; Taylor, R. T. *Polym. Sci. : Part A (poly. Chem.)* **1988**, *26*, 2651.
20. Wakselman, C.; Tordeux, M. *J. Org. Chem.* **1985**, *50*, 4047.
21. Wakselman, C.; Tordeux, M. *J. Chem. Soc., Chem. Commun.* **1984**, 793.
22. Feiring, A. E. *J. Fluorine Chem.* **1984**, *24*, 191.
23. Suzuki, H.; Yoshinaga, M.; Takaoka, K.; Hiroi, Y. *Synthesis* , **1985**, 497.
24. Volbach, W.; Ruppert, I. *Tetrahedron Lett.* **1983**, *24*, 5509.
25. Curran, D. P. *Synthesis* **1988**, 417 and 489.
26. Taylor, R. T.; Allison, S.; Green, J. W. *Polym. Prepr.* **1990**, *31*, 336.
27. Collman, J. P.; Hegedus, L.S.; Norton, J. R.; Finke, R. G. Principles and Applications of Organotransition Metal Chemistry, University Science Books: Mill Valley, California, 1987.
28. Narayagna, C.; Periasamy, M. *Synthesis* , **1985**, 253.
29. Lichstein, B. M. U. S. Patent 3,790,607; 1974.
30. (a) Haszeldine, R. N.; Kidd, J. M., *J. Chem. Soc.* **1954**, 4228. (b) Haszeldine, R. N. **J. Chem. Soc.** *1956*, 173.
31. Wakselman, C.; Tordeux, M. *Bull. Soc. Chim. Fr.*, **1986**, 868.
32. Socrates, G. Infrared Characteristic Group Frequencies, John Wiley & Sons, Ltd., New York, 1980.
33. (a) Zundel, G. *Angew. Chem. Intl. Ed.*, **1969**, *8*, 499. (b) Falk, M., *Can. J. Chem.* **1980**, *58*, 1495.
34. Hu, C. -M.; Qiu, Y. -L. *J. Org. Chem.* **1992**, *57*, 3339.
35. Some of this work has been presented. See
36. Schrauzer, G. N. *Inorg. Synth.* **1968**, *11*, 61.
37. Corey, E. J.; Schmidt, G. *Tetrahedron Lett.* **1979**, 399.
38. Silverstein, R. M.; Bassler, G. C.; Morrill, T. C. Spectrometric Identification of Organic Compounds, 5th ed., John Wiley & Sons, Inc.: New York, 1991, p 118.
39. Huang, B. -N.; Haas, A.; Lieb, M. J. Flourine Chem. 1987, 36, 49.

INORGANIC-CONTAINING AND SHAPED POLYMERS

SYNTHESIS AND STRUCTURAL CHARACTERIZATION OF INORGANIC TIN IONOMERS

Charles E. Carraher, Jr., Fengchen He and Dorothy Sterling

Florida Atlantic University
Department of Chemistry and Biochemistry
Boca Raton, FL. 33431
and
Florida Center for Environmental Studies
Northcorp Center
Palm Beach Gardens, FL. 33410
and
Motorola, Inc.
Manufacturing Research
Plantation, FL. 33322

SUMMARY/ABSTRACT

Ionomers have been synthesized from reaction of tin II and tin IV metal halides and organostannane halides through reaction with an ethylene-acrylic acid copolymer. Mass spectral, infrared spectral, and elemental analysis results are consistent with the formation of tin-containing ionomers. The products all exhibit "ionomer-like" properties and soften below 150 C, many softening below 50 C.

INTRODUCTION

The term "ionomer" is generally used to describe polymers consisting of a hydrocarbon backbone and containing 10 to 15 mole percent or less of ionizable groups such as carboxylic acid or sulfonic acid. These groups can be partially or completely neutralized by metal or corresponding counter ions.

Ionomers differ from polyelectrolytes in the proportion of ionizable groups and often in their physical properties. Polyelectrolytes generally have a higher proportion of ionizable units. Ionomers can be divided according to the distribution of the ionizable units as random, telechelic, block and segmented.

The first commercial ionomers were introduced in the mid-1960s when Dupont produced an ethylene/methacrylic acid copolymer under the trade name Surlyn (TM). In Surlyn (TM) carboxyl groups are partially neutralized with sodium and zinc cations. The structure of this ionomer can be depicted as follows for neutralization occurring through use of a monovalent cation.

$$-(-CH_2-CH_2-)_n-CH_2-\underset{\displaystyle COO^-,M^+}{\underset{|}{C(CH_3)}}-)_n$$

Today, ionomers are becoming more important as commercial materials (Table 1).

Table 1. Selected commercial ionomers

Polymer System	Tradename	Manufacturer	Application
Ethylene-methacrylic acid	Surlyn	Dupont	Modified thermoplastic
Butadiene-acrylic acid	Hycar	Goodrich	High green-strength elastomer
Perfluorosulfonates	Nafion	Dupont	Multiple membranes
Telechelic Polybutadiene	Hycar	Goodrich	Specialty
Sulfonated ethylene-propylene terpolymer	Ionic Elastomer	Uniroyal	Thermoplastic elastomer

The ionic interactions and resultant ionomer properties are dependent on the
*type of polymer backbone (plastic or elastomer)
*ionic functionality (ion content)
*type of ionic moiety (carboxylate, sulfonate, or phosphonate)
*degree of neutralization and
*type of cation (charge, form, valence)

The ionomer structure has been represented as having two general types of bonding groupings-multiple and (essentially) individual groupings (1). The mechanism of metal ion-carbonyl group interaction is usually described as an association of charged and uncharged anion-containing function units surrounding the cations. The cation is treated as point charges for Group IA and IIA metals. The maximum number of ion pairs that can interact to form a multiplet is eight (1). For the higher group metals covalent bonding contributions should also be considered where p, d and f orbitals may be involved.

The ionic clusters act as sites of cross-linking at low temperatures. The interchain forces resulting from this ionic bond produces properties normally associated with a cross-linked thermoset polymer. The association in ionomers can be partially overcome through application of heat and pressure allowing processability while "truly" cross-linked network polymers cannot be remelted, dissolved or reshaped. Thus, ionomers are often referred to as processable thermosets.

Most ionomers are transparent since the cross-linking is random, giving rise to no large crystalline regions. The toughness and clarity make ionomers an excellent glass coating material.

Since ionomers contain polar groups, their adhesion to most polar surfaces is better than that of other polyolefins. Ionomers are also employed as packaging film, golf ball covers and shoe components.

Organotin compounds are generally found in two oxidation states, tin II (stannous) and tin IV (stannic). Organotin compounds offer a wide variety of structural types in which the tin atom is two to seven coordinated (2). Geometries for organotin compounds include bent, trigonal, tetrahedral, trigonal bipyrimidal, octahedral and pentagonal bipyramidal.

Simple organotin carboxylates have been prepared using several approaches. Thus, the reaction of organotin halides with metal salts of carboxylic acids has been used to make a series of carboxylate esters (3, 4). From investigating the infrared spectra of solid and solutions it was concluded that trialkyltin carboxylates are polymeric in the solid state with planer trialkyltin groups and bridging acyloxyl groups forming the backbone linkages as noted below (5-8).

···Sn···O-CO-O···Sn···O-CO-O···Sn···O-CO-O···Sn···-O-CO-O

For diorganotin dicarboxylates, the symmetrically chelated structure has been proposed from infrared and far-infrared spectra (9). A nonsymmetrically chelated configuration and a partial bridging of acetoxy groups have also been proposed based on infrared data (10, 11).

Several groups have synthesized organotin polyesters (12-15). These polymers are believed to have the following backbone structure.

```
    R
-(-Sn-O-CO-R-CO-O-)-
    R
```

We reported the modification of poly(acrylic acid), PAA, through reaction with a number of Group IVA organo halides (for instance 16). Reaction with triaryl or trialkylmetallic halides results in the formation of pendant organometallic groups while reaction of PAA through reaction with diaryl and dialkylmetallic dihalides gives crosslinked products.

Here we report the synthesis of inorganic tin II and tin IV ionomers based on a copolymer of ethylene and acrylic acid.

EXPERIMENTAL

Material

The pre-ionomer was used as received: ethylene/acrylic acid copolymer, acrylic acid content 20%, Scientific Polymer Products, Inc., Ontario, NY;

Synthesis and Ionomer Characterization

The acid content of the ethylene/acrylic acid copolymer was determined using a potentiometric titration employing standardized sodium hydroxide.

Synthesis of the organotin-containing ionomeric materials was effected utilizing the interfacial polycondensation procedure. An aqueous solution of ethylene/acrylic acid copolymer containing sodium hydroxide was added to a one-pint Kimax emulsifying jar fitted on top of a Waring Blender (model 1120) with a no-load speed of about 18,000 rpm. The cap was screwed on and the blender turned on. An equivalence (based on the acid group present and the potential reactivity of the tin-containing unit) of the organotin-containing reactant in an organic liquid (generally 50 ml of chloroform or carbon tetrachloride) was added. After 30 seconds the stirring was stopped. Reaction generally occurred within 15 seconds based on precipitate formation.

Synthesis of the inorganic tin-containing materials was similarly effected except the tin salt was dissolved in water. Thus, the sodium ionomer containing 0.00200 moles of acrylic groups is dissolved in 50 ml of water and added to the Kimax emulsifying jar. Stirring was begun and a second solution containing 50 ml of water and 0.00050 mole of tin tetrachloride or 0.00100 mole of tin dichloride. The blender was run for 30 seconds.

The precipitate was recovered using suction filtration, washed repeatedly with distilled water and then washed into a glass petri dish and allowed to dry.

Solubility was tested by placing about 1 mg of solid in about 3 ml of liquid.

Physical Characterization

Softening ranges were obtained using a Fisher-Johns melting point apparatus. Thermogravimetric analyses, TGA, were obtained using a Model 951 Dupont Thermogravimetric Analyzer. About 2 mg of sample was used. A heating rate of 20 C/min was employed with heating from room temperature to 800 C. Differential Scanning Calorimetry, DSC, was conducted using a Dupont DSC Cell Base connected to a Dupont Model 990 Thermal Analyzer console. A Mettler H20T semimicro balance was employed to weight the samples. Heating was room temperature to about 450 C at a heating rate of 20 C/min. A gas flow rate of about 50 ml/min was employed for both DSC and TGA.

Infrared spectra were recorded using a Mattson Instruments Galaxy FTIR employing KBr pellets and films. Spectra were recorded using an instrument resolution of 4 wavenumbers using 32 scans. High resolution positive ion mass spectroscopy was carried out at the Midwest Center for Mass Spectroscopy, Lincoln, Neb. using a Kratos MS-50 mass Spectrometer. The samples were inserted as solids in a glass ampule using a direct insertion probe and ballistically heated to about 450 C.

RESULTS AND DISCUSSION

One primary reason for synthesizing ionomers from organotin halides is to test the modes of movement that are necessary for the material to exhibit ionomer-like behavior with respect to movement when heat and pressure are applied. Ionomer movement is believed to occur through a combination of simple ethylene-unit displacement and through movement of the metal-associated sites with the metals acting as "ball-bearings". For ionomer-like materials formed through reaction with organotin halides, the latter type of movement is probably disallowed because of the directional requirements of bonding between the organotin moieties and the carboxylate grouping. Even so, crosslinked products from diorganotin dichloride and monoorganotin trichloride do exhibit ionomer-like melting indicative that such behavior can result from only the movement of the ethylene segments (for instance 17, 18). By comparison, the inorganic tin ionomers reported here should be considered as "traditional" types of ionomers offering the possibility of movement through both pathways.

The ethylene acrylic acid copolymer is sold as containing 20 % (by weight) acrylic acid. Experimentally it was found to contain 22.9 % acrylic acid corresponding to about 10 % of the units being acrylic acid.

The synthetic procedures are general giving products using organotin trichlorides, diorganotin dichlorides, triorganotin chlorides and tin II and tin IV. Sample results appear in Table 2. The molar ratio of tin-containing reactant was based on the assumption that all of the active sites of the reactants would be reacted.

Table 2. Selected yield results as a function of tin-containing reactant. Two mmole of preionomer is used.

Tin-Containing Reactant (mmole)	Yield (g)	Yield (%)*
Dibutyltin dichloride (1)	0.68	79
Dilauryltin dichloride (1)	0.88	96
Tributyltin chloride (2)	1.1	9
Triphenyltin chloride (2)	0.1	1
Butyltin trichloride (0.67)	0.80	90
Methyltin trichloride (0.67)	0.74	100
Phenyltin trichloride (0.67)	0.22	96
Tin II (1)	0.26	100
Tin IV (0.5)	0.09	36

Results of tin analysis for the products are given in Table 3. Two theoretical calculations of percentage tin were made. The minimum percentage of tin was calculated assuming all of the halide groups were reacted. For instance, for dihaloorganostannanes this would have one organostannane moiety for every two acid groups.

Table 3. Tin analysis results and theoretical calculations.

Tin-Containing Reactant	%-Tin Found	%-Tin Calc.-Min	%-Tin Calc.-Max	Substitution (%)
Dibutyltin dichloride	13.9	14	22	51/102
Dilauryltin dichloride	11.7	12	16	57/114
Tributyltin chloride	7.85	20		26
Triphenyltin chloride	18.4	18		100
Butyltin trichloride	17.6	11	24	63
Methyltin trichloride	9.14	11	27	25
Phenyltin trichloride	16.4	11	23	61
Tin II	10.3	14.6		67
Tin IV	5.6	14.6		67

Such a product would be cross-linked. The maximum percentage of tin was calculated assuming that only one of the active groups of the organotin halides was reacted. Again, for dihallorganostannanes, this would have one organostannane for every acid group. In this case, only linear products would be produced. For products derived from monohalostannanes, these maximal and minimal values are the same.

The tin content for the tin II product is 10.3 % corresponding to about one tin atom for every three acrylic acid units. For the tin IV product with a tin content of 5.6 % this corresponds to one tin atom for every six acrylic acid units. These results are reasonable. The tin II product can be envisioned to "neutralize" or react with two acrylic acid units for a reaction efficiency of about 67 %; that is about 2 out of every 3 acrylic acid units would be reacted. The tin IV product can similarly be envisioned to react with four acrylic acid units for a reaction efficiency also of about 67 % with about 4 out of every 6 acrylic acid units involved in the bonding. Space considerations are probably responsible for the "loading" being less than 100 %.

The inorganic tin-containing products are insoluble in all tested liquids consistent with the behavior of other tin-containing cross-linked ionomers. Liquids tested include DMF, DMA, HMPA, DMSO, 1-methyl-2-pyrrolidine, acetonitrile and acetone.

The ionomer structure is probably a combination of reacted and unreacted sites that may be part of other "chelated" sites. Thus, it is appropriate to consider an approximate structure such as that given below where all of the acrylate and tin groups are reacted.

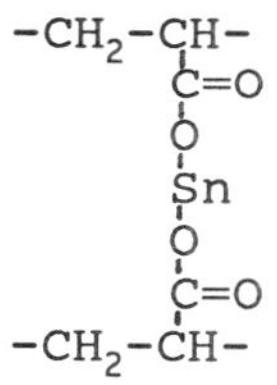

TIN II STRUCTURE

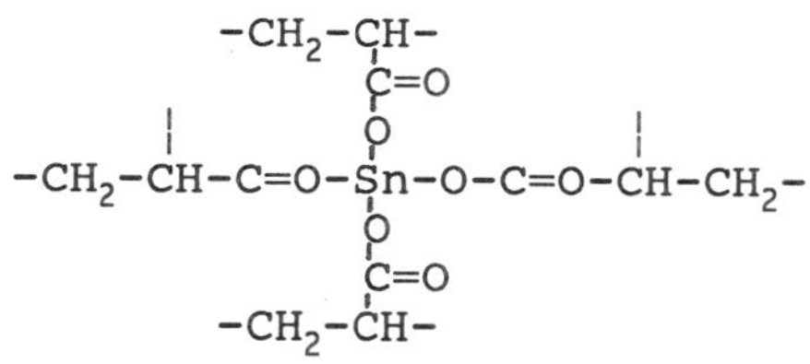

TIN IV STRUCTURE

Mass spectral results are consistent with the proposed structures and are given in Tables 4 and 5. Almost all of the major ion fragments are derived from the simple degradation of the hydrocarbon backbone. Ion fragments can be grouped together with respect to the number of carbons present within the particular ion fragment cluster and are reported as such in Tables 4 and 5.

Ion fragments consistent with the formation of the bonding between the inorganic tin and acrylic acid are present. Most of these are of low intensity, but many give isotopic abundances for the two most abundant tin isotopes in near the expected relative abundance (the abundance ratio of 120 to 118 is 1.4). Where possible, the relative abundance of 120 to 118 are cited. For the

Table 4. Major ion fragments derived from rapid heating to 450 C for the product of tin II and the preionomer.

m/e	Relative Intensity (%)	Assignment
41	68	
43	20,90	C_3H_x
44	75	CO_2
45	20	
55	24,95	C_2H_3CO
56	31	C_4H_x
57	100	
60	24	
67	36	
68	17	C_5H_x
69	83	
70	31	
71	61	
73	26	
81	43	
82	24	
83	63	C_6H_x
84	27	
85	37	
95	36	
96	20	C_7H_x
97	46	
98	37	
109	22	C_8H_x
111	20	
236	20	$Sn(COO)_2CHHCH$

product from tin IV ion fragments consistent with the formation of a single Sn-O linkage are found at (all ion fragments are given in m/e = 1) 136/134 assigned to Sn-O; 164/162 assigned to $Sn-CO_2$ (ratio of 120 to 118 = 1.2); 177/175 assigned to $SnCO_2CH$ (ration = 1.5); and 191 assigned to $SnCO_2CHHCH$. Ion fragments consistent with the formation of two chelates with the tin indicative of cross-link formation are found at 208/206 assigned to $Sn(CO_2)_2$ (ratio = 1.2); 180 assigned to $O-SnCO_2$; 219 assigned to $Sn(CO_2)_2CH$; 237/235 assigned to $Sn(CO_2)_2CHHCH$ (ratio = 1.5): this moiety will be described as "M" in the following ion fragment descriptions); 249/247 assigned to MHCH (ratio = 1.1); 263/261 assigned to MHCHHCH (ration = 1.4); 277 assigned to MHCHHCHHCH; 291$M(CH_2)_4$; 319 assigned to $M(HCH)_6$, and 333 assigned to $M(HCH)_7$. Ion fragments indicative of tri-chelation with the tin are found at 252/250 assigned to $Sn(CO_2)_3$ and 266 assigned to the former structure plus a methylene. Finally, ion fragments indicative of tetra-chelation with the tin are found at 296 assigned to $Sn(CO_2)_4$; 311 assigned to the previous ion fragment structure plus a methylene; and 325/323 assigned to the previous structure plus another methylene.

Similar ion fragments are found for the tin II product consistent with the proposed structure.

Ion fragments are absent between 35 to 38 consistent with the absence of Cl for the tin IV product. The tin II product shows a low intensity band at 36 within this range.

The infrared spectra of the products are consistent with the proposed structures. The tin IV product shows a major band

Table 5. Major ion fragments derived from the product of reaction between the preionomer and tin IV.

m/e	Relative Intensity (%)	Assignment
41	60	
43	74	C_3H_x
44	100	
45	13	
55	78	
56	30	C_4H_x
57	84	
60	12	
67	24	
68	18	
69	73	C_5H_x
70	31	
73	17	
81	30	
82	35	
83	53	C_6H_x
84	13,15	
85	29	
95	21	
96	18	
97	34	C_7H_x
98	13	
109	12	C_8H_x
111	17	

centered about 3450 (all bands given in cm^{-1}) corresponding to unreacted and un-neutralized acid groups. Bands about 2900 and 2840 correspond to C-H aliphatic stretching. Skeletal -C-C- vibrations appear within the 700 to 760 range. For the product they occur about 720. The O-H out of plane deformation occurs at about 950. For the product it occurs at about 940. The product shows a band about 1240 corresponding to the C-O stretching.

The inorganic tin atom can be connected to the carbonyl through what is commonly called bridging and non-bridging as illustrated below.

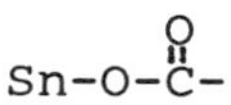

Non-bridging

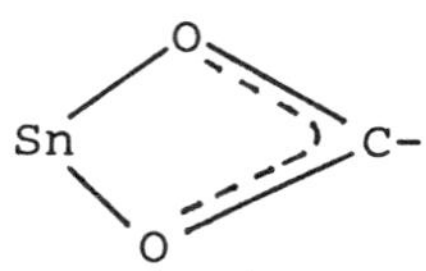

Bridging

In general, the tendency towards bridging for organotin carboxylate compounds is Pb>Sn>Ge>Si. In the infrared, bands associated with bridging metal carboxylates occur about 1570 corresponding to asymmetric stretching and a band about 1420 corresponding to the symmetric stretching. For non-bridging, the asymmetric stretching is found about 1600 to 1650 while the symmetric stretching is found about 1360. For the product derived from tin IV major bands are found at about 1610 and 1430 consistent with carboxylate bonding being of the non-bridging variety. Further, lesser bands are also found at 1570 and 1240 characteristic of bridging. Thus, both types of bonding are found with the major bonding type being of the non-bridging type. By comparison, the analogous organotin ionomers bonding is mainly of the bridging variety (17, 18). Thus, there exists a difference in the tendency between the organotin and inorganic tin-containing ionomers with respect to bonding to the carboxylate moiety. Infrared results for the tin II products are similar to those found for the tin IV products.

The tin II product shows four peaks (from DSC data)-at 60, 85, 115 and 350 C. The tin IV product shows only two peaks at 85 and 370 C. The sodium ionomer shows peaks at 60, 95, 130 and beginning at 450 C. The pre-ionomer shows peaks at about 40, 90 and 430 C. In all cases, the highest transition appears to coincide with a loss in weight observed in the TGA and the lowest (approximately) corresponds to the softening temperature observed with the Fisher-Johns apparatus. It is believed that the lowest temperature corresponds to movement in the ethylene units and that the additional peaks may correspond to more complex movements involving the carboxylate moieties.

It is possible that the reason for the lesser number of DSC peaks for the tin IV product is that it is more highly cross-linked thus restricting motion, and associated phase-change rearrangements, relative to the tin II products.

By comparison, the products derived from monohalo, dihalo and trihalo-organotin reactants exhibit initial softening (both through visual observation utilizing the Fisher-Johns and DSC) in the range of 40 to 140 C.

The ionomers typically offer good weight retention to about 300 C after which the backbone decomposes. For the inorganic tin II product 13 % weight retention is found to 750 C. Based on the tin analysis, 12.5 % SnO (or 13.9 SnO_2) should be retained assuming that no tin was evolved. For the tin IV product 12 % of the weight was retained. Again, based on the tin elemental analysis 6.3 % SnO or 7.1 % SnO_2 would be present if the tin was converted to tin oxide. The pre-ionomer itself shows a weight retention in air and nitrogen of less than 5 %. The results are consistent with the major part of the tin remaining as a solid residue.

REFERENCES

1. A. Eisenberg, Macromolecules, 3, 147 (1970).
2. I. Omae, Organotin Chemistry, Elsevier, Amsterdam, 1989.
3. H. H. Anderson, Inorganic Chem., 3, 912 (1964).
4. N. Zemlyanskii, E. Panov, and K. Kocheshkov, Zhur. Obshchei. Khim., 32, 291 (1962).
5. M. Janssen, J. Luijten and G. van der Kerk, Rec. Trav. Chim., 82, 90 (1963).
6. R. Cummins and P. Dunn, Australian J. Chem., 17, 185 (1963).
7. R. Okawara and M. Ohara, Bull. Chem. Soc. Japan, 36, 624 (1963).
8. R. Okawara and M. Ohara, J. Organometallic Chem., 1, 360 (1964).
9. H. Sato and R. Okawara, Int. Symp. Mol. Struct. Spectry., Tokyo, Japan, September, 1962.
10. Y. Maeda, C. Dillard and R. Okawara, Inorg. Nucl. Chem. Letters, 2, 197 (1966).
11. Y. Maeda and R. Okawara, J. Organometallic Chem., 10, 247 (1967).
12. M. Frankel, J. Applied Polymer Sci., 9, 3383 (1965).

13. C. Carraher and R. Dammier, J. Polymer Sci. A-1, 8, 3367 (1970).
14. C. Carraher, Makromolek. Chem., 135, 107 (1970).
15. S. Migdal, D. Gerther and A. Zilkha, J. Organometallic Chem, 11, 441 (1968).
16. C. Carraher and J. Piersma, J. Applied Polymer Sci., 16, 1851 (1972).
17. C. Carraher, F. He and D. Sterling, PMSE, 72, 114 (1990).
18. C. Carraher, F. He and D. Sterling, Synthesis, Characterization and Theory of Polymeric Networks and Gels, S. M. Aharoni, Ed., Plenum, NY, 1992.

COMPUTER MODELING OF POLY(ACRYLIC ACID) AND ITS SALTS

Xinhua Xu, Charles E. Carraher, Jr. and Mark D. Jackson
Florida Atlantic University
Department of Chemistry and Biochemistry
Boca Raton, FL 33431

Florida Center for Environmental Studies
NorthCorp
Palm Beach Gardens, FL 33410

SUMMARY

Poly(acrylic acid), PAA, and its salts were modeled using PCMODEL and MNDO. The results agree with research results. The modeling results are consistent site of cation "attack" is the hydroxylic oxygens. These hydroxylic oxygens radiate away-from the hydrocarbon central core making them particularly accessible for substitution reactions. PAA and salts containing sodium and magnesium atoms approximate helical structures.

INTRODUCTION

Computer modeling is becoming more important to the bench chemist in allowing the chemist a way to "picture" what might be happening. There are many computer modeling programs presently available, many offering selected advantages over their rivals. Here we describe, only briefly, the results of one of these studies as they relate to other studies contained in this book.

Poly(acrylic acid) is water soluble. Because of its water solubility and its ability to increase the viscosity of water, it is used as a thickener. It is also a good flocculent for sewage treatment and is added as a pigment dispersant in latex paints, and is used in binders and adhesives. Polymers and copolymers containing acrylic or/and methacrylic acid are manufactured at a rate of about 2,000,000 metric tons yearly.

Neutralization of PAA leads to the formation of a polyelectrolyte. The Henderson-Hasselbalch equation can be used to relate the pH of a solution and the extent of neutralization. The Henderson-Hasselbalch equation is modified for polyelectrolytes since the negative charge concentration on the polymer chain is increased by increasing neutralization. Further, electrostatic work is then required to remove a proton from the polyelectrolyte to infinity. Thus, the increased local negative group concentration decreases the acid-moiety dissociation giving a decreased acid strength. The pH of propenoic acid itself is 4.25 whereas the pH of PAA is 4.75 (1).

For the neutralization reaction

$$HA + BOH \longrightarrow HOH + A^- + B^+$$

the fraction of acid groups neutralized, or the degree of neutralization, can be defined as

$$f = [HA]/([HA] + [A^-])$$

Along with the electrostatic effect, a statistical or entropy effect is also present and accounts for the increased dependence of the equilibrium constant on the degree of neutralization, f, with decreased acid strength of the polyelectrolyte. Thus, cations from the base compete with protons from the acid for positions on the PAA chain. There are more possibilities or opportunities for a cation with respect to replacing a proton when f is below 0.5 (that is when neutralization is less than one half). The counter is true when f is above 0.5. Thus, PAA is a stronger acid compared with monoacids such as propenoic acid at low fractions of neutralization, whereas PAA is a weaker acid at high fractions of neutralizations. However, this entropy effect is overwhelmed for PAA by the electrostatic effect with the end result that PAA is a weaker acid than propenoic acid throughout its neutralization.

In fact, concentrated solutions of sodium chloride are used to precipitate partially neutralized PAA (2). Further, if partially neutralized, even dilute solutions of PAA can be precipitated through addition of alkaline earth salts (3).

Titration curves of PAA rise only slowly with increasing neutralization. Because of this, partial salts of PAA are good buffers in the pH range of 4 to 6.4. PAA is difficult to titrate because equilibrium is only slowly established (4).

The acidity of PAA is also dependant on the size of the cation (5). For instance, the smaller size of the lithium cation allows it to act more like a point charge and to be more tightly bound to the acid site. It is said to have a higher surface charge density. By comparison, the larger rubidium cation is less firmly held by the acid group. Thus, PAA appears to act as a stronger acid when neutralized by lithium hydroxide compared with neutralization by addition of rubidium hydroxide. Often, additional time and amount of cation is needed for large cations. This is probably a consequence of both the electrostatic charge effect and steric considerations. Thus, neutralization by magnesium and the zirconyl, presented in the next paper, will occur readily, not because of the size, but rather because of the higher charge of the cations.

PAA can be prepared using bulk polymerization, aqueous polymerization, nonaqueous polymerization, inverse phase emulsion and suspension polymerization. The precise structure of the resulting PAA chain is dependent upon many factors including the polymerization process and conditions. The tacticity of poly(methacrylic acid), PMA, has been studied using NMR spectroscopy (6). For polymerization of methacrylic acid in methyl ethyl ketone at 60 C gives a polymer with 57% syndiotactic triads. Polymerization at low temperatures gives a more syndiotactic product (6) as does polymerization at high pH (7).

EXPERIMENTAL

For the present study, only a six-repeat unit PAA (or PAA-derived structure) was used. Two modeling programs were employed-PCMODEL and MNDO (8-10). In general, the model is asked to create the most stable structures using given bond angles and for different atoms situated on the carboxylate moiety. Literature values for the various bond lengths were used-namely C=O 1.208 A; C-O 1.342 A; O-H 0.972 A; C-H 1.100 to 1.110 A; and C-C 1.519 to 1.545 A.

RESULTS AND DISCUSSION

Molecular modeling serves many purposes. Here we are interested in an optimization of the geometry of specific molecules given known bond lengths, bond angles and energy data. The goal is to find a structure most consistent with ideal structural features. Each type of atom, and each kind of bond must be parameterized from experimental data.

Semi-empirical molecular orbital, MO, theory uses a combination of experimental data and quantum mechanical MO methods to model the valence electronic structure of molecules. In the MNDO (8) method each atom is parameterized using experimental data. This calculation provides molecular orbital descriptions of the valence electrons, as well as effective charges of each atom in the molecule.

In this study, the molecular modeling calculations are carried out using the PCModel molecular modeling software package (9). Optimized geometries are based on the MMX force field, a variation of the MM2 model of Allinger (10). The energies obtained are strain energies, that is, the energy relative to an ideal bonding configuration.

The electronic charges on the respective atoms are determined using fixed-geometry MNDO calculations. The geometries for these calculations are obtained from the PCModel studies.

Propenoic (acrylic acid) units can be added in three major sequences. These sequences are random (corresponding to an atactic arrangement of the acid groups), trans (corresponding to a syndiotactic arrangement) and cis (corresponding to an isotactic arrangement). For the present study, we looked at only the latter two arrangements. These two arrangements yield two different structures. The "trans" sequence will give a "drawn-out" chain

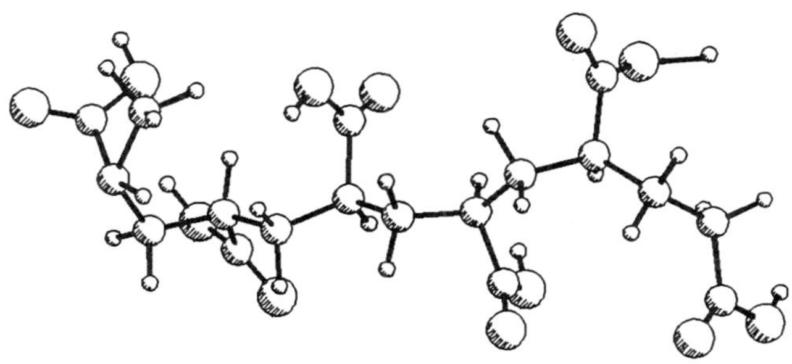

Figure 1. Optimized "trans" (syndiotactic) polymer.

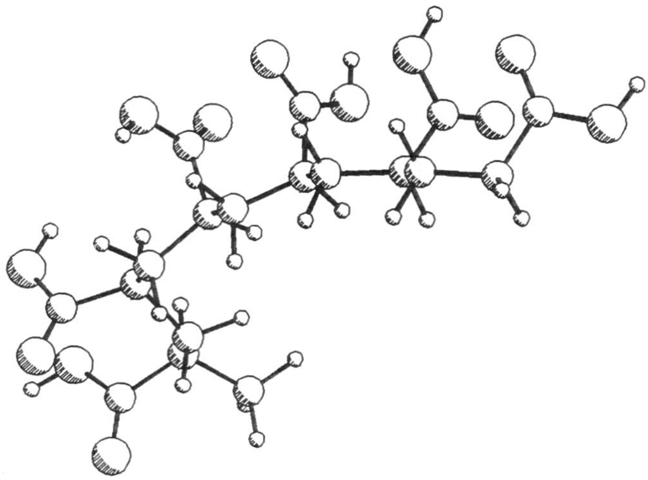

Figure 2. Optimized "cis" (isotactic) polymer.

structure whereas the "cis" sequence allows the formation of a helix. The minimized energy for the "trans" form is 112.87 whereas the minimized energy for the "cis" arrangement is 24.05. "Minimized" energy structures are given in Figures 1 and 2. It is seen that the "cis" structure is helical-like. Thus, the helix structure is favored and probably occurs because of the formation of internal hydrogen bonding. Even so, helical formation for most polymers is favored, when possible, because this structure allows the like units composing the polymer backbone, here hydrocarbon units, to associate. In water, such associations should be even more strongly favored in comparison to solution in less polar solvents.

Strong association of the carboxylate moieties through hydrogen bonding occurs in aqueous solutions as well as in dry and

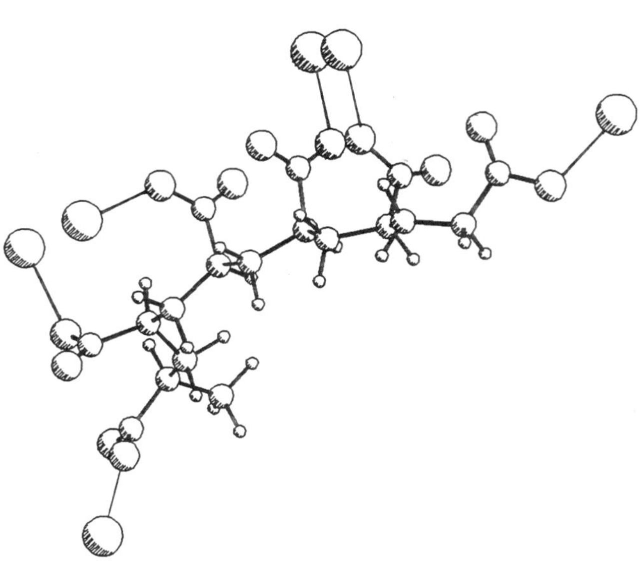

Figure 3. Optimized isotactic poly(sodium acrylate).

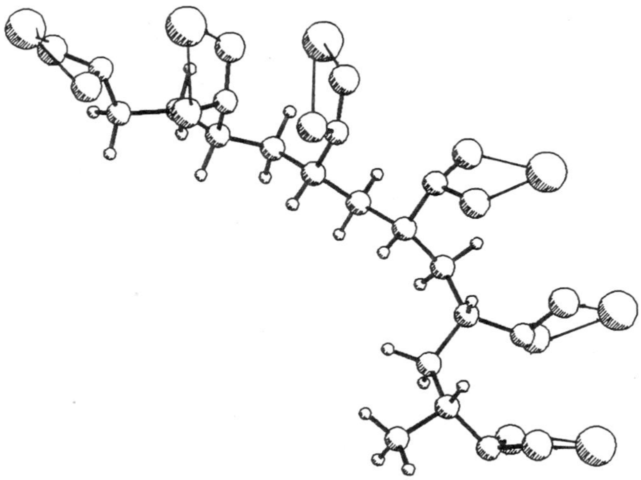

Figure 4. Optimized isotactic poly(magnesium acrylate).

hydrated forms (11). This is consistent with what the model shows.

The next calculation done was with ("cis") PAA fully neutralized through reaction with sodium hydroxide-that is sodium cations replacing the protons on the carboxylate moiety. Here, the non-bridging structure was assumed for calculations sake alone. The bond lengths between the sodium and oxygen, after energy minimization, have increased from a value of about 0.972 A to 2.247-2.393 A. This is mainly due to the larger size of sodium compared to that of hydrogen. A minimization energy of 128-units? was found. The somewhat helical structure persists (Figure 3). It is important to note that the carboxylate groups are "pointing-out" available for substitution reactions to occur.

Next, calculations were done on ("cis") PAA fully neutralized using Mg in place of H. Here, a bridging structure was assumed between the Mg ion and a single carboxylate unit. (Because the Mg ion is doubly charged, and each carboxylate unit is singly charged, the Mg ion will end up with a full negative charge that can be satisfied through reaction with a second carboxylate unit.) Again, the somewhat helical structure is present (Figure 4). The Mg-O bond length is 2.110-2.253 after energy minimization. Thus the Mg-O bond length is less than that of Na-O. This is due to both the ionic radii of Mg (86 picometers) being smaller than Na (116 picometers) and that the sodium cation has a single charge while the Mg is doubly charged. The minimization energy for this structure is 215.

The net charge on the various atoms were calculated. All of the oxygen atoms have negative charges (-0.29 to -0.36) while all of the hydrogens have positive charges (0.22 to 0.24). This is consistent with the acid groups being the reactive sites with cations.

In summary, modeling results are consistent with experimental results with respect to molecular shape and the reaction site.

REFERENCES

1. R. H. Wiley and G. M. Braver, J. Polymer Sci., 3, 647 (1948).
2. I. Williamson, Br. Plast., 23, 87 (1950).

3. ASTM Standards, ASTM D 1043-61T, Vol. 27, American Society for Testing Materials, Philadelphia, PA, 1964.
4. A. V. Tobolksy, J. Polymer Sci., C9, 157 (1975).
5. US Pat. 2,932,749 (April 5, 1960), B. B. Kine and N. A. Matlin (to Rhome and Haas Co., Philadelphia, PA).
6. D. W. Van Krevelen, Properties of Polymers, Elsevier, Amsterdam, 1976.
7. J. Brandrup and E. H. Immergut, Polymer Handbook, 2nd. Ed., Wiley-Interscience, NY, 1975.
8. M. J. S. Dewar and W. Thiel, J. Amer. Chem. Soc., 99, 4899 (1977).
9. PCModel Molecular Modeling Software, Serena Software, Bloomington, IN, 1990.
10. U. Burket and N. L. Allinger, Molecular Mechanics, ACS Monograph 177, American Chemical Society, Washington, D. C., 1982.
11. J. E. Nemec and W. Bauer, Encyclopedia of Polymer Science and Engineering, 2nd Edition (H. Mark, N. Bikales, C. Overberger and G. Menges, Eds.), Vol. 1, Wiley, NY, 1985.

SYNTHESIS AND STRUCTURAL CHARACTERIZATION OF TITANOCENE-CONTAINING POLYETHERS BASED ON REACTION WITH ETHYLENE OXIDE-CONTAINING DIOLS, INCLUDING POLY(ETHYLENE GLYCOL)

Charles E. Carraher, Jr. and Lisa Reckleben

Department of Chemistry and Biochemistry
Florida Atlantic University
Boca Raton, FL 33431 and

Florida Center for Environmental Studies
Palm Beach Gardens, FL 33410

SUMMARY/ABSTRACT

Many metal-containing polymers are only difficulty soluble or are insoluble, exist only as powders, and degrade without softening making processing difficult. In an effort to overcome these difficulties a series of titanocene polyethers were synthesized. The products were formed through reaction of hydroxyl-capped poly(ethylene glycol)s with titanocene dichloride. While some of the products softened (PEG> 1000 Daltons) in the 28 to 55 C range, none were soluble even utilizing heated solutions.

INTRODUCTION

Many metal-containing polymers do not melt prior to thermal degradation and they do not dissolve even with heating (for instance 1-4). A number of factors appear to be responsible for these problems. These factors include

*low degree of flexibility
*unique combination of hydrophobic and hydrophilic structures and
*high degree of crystallinity.

This lack of solubility and softening makes processing difficult. In an effort to over come these difficulties a number of scenarios have been considered including the following.

*copolymerization

*use of flexibilizing units
*use of non-symmetrical reactants and
*use of large substitutes.

Copolymerization can be effected through either varying the Lewis base or Lewis acid. We have had only moderate success with this approach. We have also had only limited success using large substitutes such as the tert-butyl group and working with reactants that have differing Lewis base reactive sites such as amino acids. We have had better success combining features. Thus, ricinoleic acid, derived from castor oil, has both structural dissymmetry, Lewis base dissymmetry (one alcohol and one acid Lewis base) and a number of "flexibiliging methylene" units. The products from group IV B metallocene dichlorides are soluble in a range of dipolar aprotic liquids, fibers and films can be drawn and cast, etc. (5).

$$CH_3(CH_2)_5CHOHCH_2CH=CH(CH_2)_7COOH$$

Ricinoleic Acid

In general the presence of certain chemical groups are known to introduce stiffness into molecules. These groups include aromatic moieties, carbonyls, amines and amides. Other groups are believed to introduce flexibility into compounds. In general, the flexibilizing character is believed to increase as follows

methylene<ethers (such as ethylene oxide units)<dimethylsiloxane

We have previously synthesized a number of titanium polyethers of the following general structure (3,4). A variety of aromatic and aliphatic diols were successfully incorporated as the fundamental backbone units of the polyethers. These materials are insoluble and do not soften prior to thermal-induced degradation.

$$-[-(Cp)_2Ti-O-R-O-]-$$

Here we report the results of using a series of hydroxyl-capped poly(ethylene glycols), PEGs. These PEG can be considered to be of the form below.

$$HO-(-CH_2-CH_2-O-)_n-H$$

Dependant on the molecular weights, PEGs range from viscous fluids to wax-like solids, with the tendency towards becoming solid increasing as the molecular weight increases (6). All of the PEGs are water soluble (6).

EXPERIMENTAL

Chemicals

Unless noted otherwise, the reactants were obtained from Aldrich Chemical Company, Milwaukee, Wisconsin and were used as received. The ethylene oxide-dimethylsiloxane-ethylene oxide ABA block copolymer was obtained from Petrarch Systems, Inc., Bristol, PA.

Polymerization Procedure

The interfacial condensation technique was used to synthesize the products. The reaction was conducted using a one quart Kimax emulsifying jar placed on a Waring Blender (Model 1120; a no-load speed of about 18,000 rpm). An aqueous solution of the diol (1.00 mmole in 50.0 ml distilled water) was prepared and sodium hydroxide (2.00 mmole) added. This solution was added to the reaction vessel and stirring begun. An organic solution containing titanocene dichloride (1.00 mmole in 50.0 ml of chloroform) was rapidly (less than 5 seconds) added to the reaction vessel through a powder funnel. Timing was begun after the completion of the addition of the organic phase. The reaction mixture was allowed to stir for 60 seconds (though the reaction was probably completed prior to this). A yellow to brown precipitate was recovered by vacuum filtration. The product was washed with several volumes of water and chloroform to remove unreacted materials. The product was then washed with water into a glass petri dish and left to dry in the air at room temperature.

Characterization

Solubility Tests

Solubilization of the products was attempted by placing about one milligram of product in three milliliter of liquid and observing over a week. The mixtures were then heated to seventy five degrees celsius in a water bath for two days. The liquids tested included DMSO, DMF, DMA, acetone, 1-methyl-2-pyrrolidinone, diethylether, p-dioxane and water.

Elemental Analysis

Titanium analyses were carried out through a combination of heating and use of perchloric acid until the residue remained a constant weight and the residue was white indicative of conversion of the titanium to titanium dioxide.

Spectral Analysis

High resolution electron impact mass spectral analyses were carried out at the Midwest Center for Mass Spectrometry, University of Nebraska, Lincoln, Nebraska. The samples were rapidly heated to about 450 C.

Infrared spectra were recorded using a Mattson Galaxy (Model 4020) FTIR. All spectra were recorded at an instrumental resolution of 4 cm^{-1} using 32 scans. The products were incorporated into KBr pellets. Additional spectra were obtained using the Spectra-Tech (Model Baseline) diffuse reflectance accessory with a sodium chloride matrix.

Thermal Analysis

Softening ranges were determined using a Fisher Mel-Temp apparatus equipped with a thermometer calibrated to 300 C.

A Dupont Thermogravimetric Analyzer was used to obtains thermograms of the products. A heating rate of 20 degrees per minute and a gas flow rate of 70 ml/min was used.

RESULTS AND DISCUSSION

A series of titanium polyethers (repeat unit given below) based on PEG were synthesized (Table 1). The percentage yields were, as expected, similar except for the siloxane ABA block copolymer. In general, there was a slight decrease in percentage yield as the molecular weight of the PEG increased possibly due to size-related factors and the ability of the reactive sites to "get together".

$$-(-Cp_2Ti-(CH_2-CH_2-O-)_n-)-$$

Table 1. Percentage yields as a function of PEG. (Reaction conditions: One millimole of titanocene dichloride in 50 ml of chloroform added to a stirred solution of one millimole of diol in 50 ml of water.)

PEG-MW	%-Yield
200	68
400	42
600	55
1000	51
1500	52
2000	49
3400	46
8000	41
ABA-Block Copolymer	17

Elemental analysis for titanium was in mild agreement with the calculated. For instance for the product from PEG-MW2000, the calculated percentage of titanium is 2.2 %; found was 2.9; and for the product from PEG-MW3400 is calculated to be 1.3 % and 1.6 % was found.

Infrared Spectra

In general the infrared spectra is in agreement with the proposed structure. Table has the infrared spectral band assignments for titanocene dichloride. Table 2 contains the band assignments for PEG-MW200 and the product. All bands are given in 1/cm. For PEG-associated bands, little shift is found, as expected, since the bulk of the non-titanocene ethylene oxide are found in an environment similar to that of PEG itself. The band assignments given in Table 2 are consistent with those reported by Balducci et al. (7). Band assignments from titanocene-containing products arising from the Cp moiety should be simplified since the frequency of these vibrations are reported to remain relatively constant upon substitution of the halide ligands (7-10).

The infrared spectra for high molecular weight PEGs (MW>2000) have been studied in both the crystalline and molten states (11-13). Assignments given in Table 3 are in agreement with those reported elsewhere (11-13).

Thus, it is expected that the infrared spectra of the products should be similar to a combination of the two reactants. This is what is found (for instance Table 3). The O-H stretch at about 3400 is less, again as expected.

Table 2. Band assignments for titanocene dichloride.

Band Assignments	Frequency
C-H stretch	3107
C-C stretch	1440, 1371
C-H i.p. bend	1014
C-H o.p. bend	873, 827

Table 3. Band assignments for PEG-MW200 and its product.

Band Assignments	PEG-MW200	Product
C-H stretch	2887	2889
HCH scissor	1469	1471
HCH wag	1415, 1346	1411, 1342
HCH twist	1282, 1242	1282, 1242
C-O stretch, HCH rock	1143	1145
C-C, C-O stretch	1112	1112
C-C, C-O stretch, HCH rock	1058	1062
HCH rock, HCH twist	962	964
HCH rock, C-C stretch	949	949
C-C, C-O stretch, HCH rock	842	846

The infrared spectra of the ethylene oxide-dimethylsiloxane-ethylene oxide ABA block copolymer was also studied. Band assignments are based on the spectra of the reactants and previous work (6, 14). Following are some assignments for the product. Bands corresponding to the Si-O stretching mode occur at 1090. The Si-R stretching and rocking modes occur at about 1257 and 806 respectively. The Si-O, C-O stretch appears at 1093.

Mass Spectroscopy

Mass spectral results are consistent with the proposed structure. The mass spectra of group IV B metallocene dichlorides including titanocene dichloride has been studied (15-17). Titanocene dichloride fragments by competitive and successive loss of the Cp ring and chloride ligand giving Cp_2TiCl^+ and $CpTiCl_2^+$ as the most abundant ions (15). Cyclopentadienyl ring fragments are present at (all values given in m/e = 1) 66 (Cp+H), 65 (Cp), and 39. As expected, the ion fragments associated with the titanocene moiety become less prominent as the PEG chain increases. Table 4 contains ion fragments for the product from the product of hexaethylene glycol and titanocene dichloride showing the titanium-containing ion fragments.

Lee and Sedgwick (18) studied the mass spectroscopy of PEG-MW600. The spectrum showed a repetitive fragmentation pattern every 44 units which corresponds to the molecular weight of a single ethylene oxide unit. The ion fragmentation for the higher molecular weight PEG products are consistent with the results reported by Lee and Sedgwich. Table 5 contains results for the PEG-MW1000 product. While there are some ion fragments consistent with the presence of the titanocene moiety, they are low in intensity.

Table 4. Ion fragments derived from the reaction of hexaethylene glycol and titanocene dichloride.

m/e	%-Relative Abundance	Assignment
PEG-DERIVED ION FRAGMENTS (most abundant)		
43	8	C_2H_3O
44	100	HCH-HCH-O
45	16	HCH-HCH-OH
55	14	C_3H_3O
57	19	HCH-HCH-O-CH
59	7	HCH-HCH-O-Me
69	13	C_4H_5O
71	11	HCH-HCH-O-CH-HCH
73	9	$C_3H_5O_2$
89	7	$(HCH-HCH-O)_2H$
115	7	$(HCH-HCH-O)_2CH-HCH$
246	12	$(HCH-HCH-O)_6$ -HOH
TITANIUM-CONTAINING ION FRAGMENTS		
113	2	CpTi
129	0.7	CpTiO
193	0.7	Cp_2TiO
207	1.5	Cp_2TiOCH

Table 5. Ion fragments for the product of PEG-MW1000 and titanocene dichloride.

43	31	C_2H_3O
44	33	HCH-HCH-O
45	100	HCH-HCH-OH
58	18	HCH-HCH-O-HCH
59	18	HCH-HCH-O-Me
72	10	HCH-HCH-O-HCH-HCH
73	15	O-HCH-HCH-O-CH
87	25	HC-HCH-O-HCH-HCH-O
88	20	$(HCH-HCH-O)_2$
89	33	$(HCH-HCH-O)_2H$
133	13	$(HCH-HCH-O)_3H$

Thermal Analysis

Thermogravimetric analysis, TGA, of the products were obtained over the temperature range of about 25 to 750 C. For the high molecular weight PEGs, initial weight loss occurs about 175 C in air and 250 C in nitrogen where upon a large weight loss occurs (>95%). These TGAs are (probably) similar to those of the PEGs themselves. For lower molecular weight PEGs there is a noticeable loss in weight beginning about 50 C probably corresponding to loss of some of the Cp groups (for instance 19). For the diethylene glycol product about 80 % of the weight is retained to 750 C in both air and nitrogen consistent with the high weight retention observed for other titanium polyether products (for instance 19).

The products from low molecular weight PEGs (molecular weight

<600) gave powders when ground and did not soften to 300 C. By comparison, higher molecular weight PEGs (>1000 Daltons) showed some toughness and flexibility and melted below 100 C, generally below 60 C (Table 6).

Table 6. Softening points for selected PEG-derived products.

PEG-MW	Appearance	Softening Point, C
1000	Tacky solid	28
1500	Waxy solid	36
2000	Crystalline Powder	39
3400	Crystalline Powder	49
8000	Crystalline Powder	55

Thus, while we were unable to produce soluble titanocene-containing polyethers, PEG products that do soften were synthesized.

REFERENCES

1. C. U. Pittman and C. E. Carraher in Applications of Polymers, R. Seymour and H. Mark, Eds., Plenum Press, NY, 1986.
2. C. Carraher, Chem. Tech., 43, 741 (1972).
3. C. Carraher and S. Bajah, Polymer, 14, 42 (1973).
4. C. Carraher, L. Jambaya and S. Bajah, Polymer Preprints, 18(2), 403 (1977).
5. C. Carraher and E. Frankel, PMSE, 73, 396 (1995).
6. F. Bailey and J. Koleske, Poly(Ethylene Oxide), Academic Press, NY, 1976.
7. G. Balducci, L. Bencivenni, G. Derosa, R. Gigli, B. Martini, and S. Cesaro, J. Molecular Structure, 64, 163 (1980).
8. H. Fritz, Adv. Organomet. Chem., 1, 262 (1964).
9. E. Maslowsky, Vibrational Spectra of Organometallic Chemistry, Wiley, NY, 1973.
10. P. Druce, B. Kingston, M. Lappert, T. Spalding, and R. Srivastava, J. Amer. Chem. Soc., A, 2106 (1969).
11. H. Matsuura and T. Miyazawa, Bull. Chem. Soc. Japan, 41, 1798 (1968).

12. H. Matsuura and T. Miyazawa, J. Poly. Sci.,A2, 7, 1735 (1969).
13. J. Koenig and A. Angood, J. Poly. Sci., A2, 8, 1787 (1970).
14. G. Socrates, Infrared Characteristic Group Frequencies, Wiley, NY, 1980.
15. J. Dillard and R. Kiser, J. Organometal. Chem., 16 259 (1969).
16. R. Kiser in, Characterization of Organometallic Compounds, Vol. 1, M. Tsutsoi, Ed., Interscience, NY, 1961.
17. M. Litzow and T. Spalding, Mass Spectrometry of Inorganic and Organometallic Compounds, Elsevier, NY, 1973.
18. A. Lee and R. Sedgwick, J. Poly. Sci., 16, 685 (1978).
19. C. Carraher, H. M. Molloy, T. Tiernan, M. L. Taylor and J. Schroeder, J. Macromol. Sci.-Chem., A16, 195 (1981).

SYNTHESIS OF NOVEL SILANOL POLYMERS AND COPOLYMERS BY A SELECTIVE OXIDATION OF SI—H BOND FROM CORRESPONDING PRECURSOR POLYMERS

Eli M. Pearce,* T. K. Kwei and Shaoxiang Lu

Department of Chemical Engineering, Chemistry and Materials Science
and The Herman F. Mark Polymer Research Institute
Polytechnic University
6 Metrotech Center
Brooklyn, New York 11201

ABSTRACT

A series of novel styrene- and siloxane-based silanol polymers and copolymers were synthesized by a selective oxidation of the Si—H bond with a dimethyldioxirane solution in acetone from corresponding precursor polymers. The conversion of the Si—H to Si—OH in the polymer modification proceeded rapidly and selectively. The silanol polymers obtained in *situ* showed no tendency for self-condensation to form siloxane crosslinks in solution. Moreover, stable silanol polymers in the solid states were obtained by placing bulky substitute groups bonded directly to the silicon atom. It was found that the properties of these novel silanol polymers and copolymers depended largely on substituents bonded directly to the silicon atom and silanol composition in the copolymers as well.

INTRODUCTION

Silanol functional groups have long been known as reactive intermediates in the synthesis of silicones and for use in the sol-gel process. Generally, silanols are formed by the hydrolysis of silanes with various silicon functional groups, such as halosilanes, alkoxysilanes, etc., and then transformed into siloxanes by spontaneous and catalytic condensation of the silanols.[1] Organosilanols are also known as stronger hydrogen bond donors than the corresponding alcohol analogues.[2-4] Considering the potential applications of the silanol dual functionality either as a strong hydrogen bond donor or as a chemical crosslinking agent for preparing miscible polymer blends or semi-interpenetrating polymer networks (semi-IPNs), we initiated the research on the synthesis of silanol polymers.

Polymer modification is of particular interest when the desired polymer is not readily available from its corresponding monomer by conventional polymerization methods. The primary challenge of polymer modification is to achieve a high conversion and selective modification of the appropriate functional group. In this paper, we describe a new convenient polymer modification to prepare novel silanol polymers by a rapid and selective oxidation of the Si—H bond with a dimethyldioxirane solution in acetone from their corresponding precursor polymers.

EXPERIMENTAL

Materials

4-Chlorostyrene, magnesium turnings, 1,2-dibromoethane, potassium peroxomonosulfate (Oxone, $2KHSO_5 \cdot KHSO_4 \cdot K_2SO_4$), sodium hydrogen carbonate, anhydrous tetrahydrofuran (THF), phenyl methyl sulfide (all obtained from Aldrich Co.), dimethylchlorosilane, methylphenylchlorosilane and diphenylchlorosilane (Hüls America Co.) were used as received. Benzene, styrene, methyl ethyl ketone and acetone were distilled prior to use. Twice-distilled water was used in the preparation of dimethyldioxirane solutions. 2,2'-azobisisobutyronitrile (AIBN) was recrystallized from methanol. Poly(methylhydrosiloxane) (PMHS) end capped with trimethylsilyl groups, with a reported molecular weight of 2270, was used as received (Hüls America Co).

Preparation of 4-Vinylphenyldialkyl/arylsilane

To a three-necked, round-bottomed flask containing magnesium turnings (0.2 mole) in dry THF (20 ml), about 5 ml of the solution of 4-chlorostyrene (0.1 mole) in 50 ml THF was added with a good agitation under nitrogen, followed by an addition of 1 ml 1,2-dibromoethane in 5 ml THF. After the temperature of the reaction mixture had risen to about 50 °C, the rest of the 4-chlorostyrene solution was added dropwise. The reaction mixture was refluxed for 30 min. and then cooled to room temperature. Dialkyl/arylchlorosilane (0.1 mole) in 25 ml THF was added slowly and reaction was allowed to proceed for 1 h. at room temperature. After decomposition of excess reagents with ice water, the reaction mixture was filtered and extracted with ether, dried with magnesium sulfate overnight and distilled: 4-Vinylphenyldimethylsilane was collected at 46 °C/0.7 mm Hg with a yield of 62 %. 4-Vinylphenylmethylphenylsilane was collected at 84 ~ 85 °C/0.01 mm Hg with a yield of 59 %. 4-Vinylphenyldiphenylsilane was purified by passing through a column with a cyclohexane/methylene chloride mixed solvent.

Homopolymerization of 4-Vinylphenyldialkyl/arylsilane

4 g 4-vinylphenyldialkyl/arylsilane in 5 ml benzene was introduced into an ampule of 20 ml capacity in the presence of 2,2'-azobisisobutyronitrile (AIBN) (0.2 mole %). The ampule was thoroughly degassed by three freeze-thaw cycles, sealed in argon atmosphere and then polymerized at 60 °C for 24 h. The resulting polymer was dissolved in methylene chloride and precipitated into methanol twice. A White polymer was obtained with yield great than 75 % after dried under vacuum at 40 °C for 24 h.

Copolymerization of 4-Vinylphenyldialkyl/arylsilane With Styrene

About 20 g of a mixture of 4-vinylphenyldialkyl/arylsilane and styrene was introduced into an ampule of 25 ml capacity in the presence of AIBN (0.2 mole%). The polymerization procedure was the same as that used for homopolymerization. However, the conversion was limited to about 10% (about 4 h). The copolymer was isolated by precipitation of the methylene chloride solution into methanol and dried under vacuum at 40 °C for 6 h.

Preparation of a Dimethyldioxirane Solution in Acetone

A 500 ml three-necked round-bottomed flask, containing a mixture of twice-distilled water (50 ml), acetone (40 ml), sodium hydrogen carbonate (24 g) and a magnetic stirring bar, was equipped with a gas inlet tube extending into the reaction mixture and an air condenser loosely packed with glass wool. The exit of the air condenser was connected to a condenser filled with dry ice-isopropanol. The bottom of the dry ice condenser was attached in succession to a receiving flask (50 ml) and two cold traps, all being kept in dry ice-acetone baths. Argon gas was gently passed through the reaction vessel. While applying a slight vacuum (180~220 Torr), 50 g potassium peroxomonosulfate (Oxone) was added quickly in one portion with vigorous stirring at room temperature. The slightly yellow solution (ca. 30 ml) of dimethyldioxirane in acetone (0.06 ~ 0.08 M) was collected in a receiving flask during a 45 minute period.

Polymer Modification of the ≡Si—H Containing Precursor Polymers With Dimethyldioxirane

To a methyl ethyl ketone solution of ≡Si—H containing polymers or copolymers, a cold solution (ca. -10 °C) of dimethyldioxirane in acetone was quickly added and reacted for 30 min at 0 °C. The mole ratio of dioxirane to polymer was ca. 1.2~1.3. The resulting silanol polymers or copolymers were obtained either in solution or precipitated into hexanes followed by vacuum dry at 40 °C for 24 h.

RESULTS AND DISCUSSION

The conventional methods for the synthesis of organosilanols can be accomplished by the hydrolysis of the appropriate substituted silane in the presence of catalysts such as an acid or a base.[1] This synthetic route, however, had some difficulty when applied to the synthesis of silanol polymers which demanded not only high conversion of the functional groups for polymer modification but also resistance to the transformation of silanols to siloxane by self- or catalytic condensation during the preparation.

Dioxiranes constitute a new class of organic peroxides that possess great potential as oxidants with a variety of applications in synthetic organic chemistry.[5-7] A new convenient route for the synthesis of silanol polymers has been developed by the selective oxidation of ≡Si—H bonds with dimethyldioxirane. A series of styrene-based silanol polymers and copolymers were synthesized (Scheme 1).[8,9] The precursor polymers and styrene copolymers containing ≡Si—H bond were first synthesized by free radical polymerization of the corresponding monomers or copolymerization of the monomer with styrene. The oxidation of the silane precursor polymers was carried out by an addition of a dimethyldioxirane solution in acetone. This reaction resulted in a rapid and selective conversion of the ≡Si—H to ≡Si—OH. The transformation of the ≡Si—H to ≡Si—OH bonds is shown by the infrared spectra (Figure 1) and complete disappearance of the ≡Si—H stretching vibration band at 2118 cm^{-1} and the appearance of a new broad band centered at 3367 cm^{-1} with a weak shoulder at 3617 cm^{-1} assigned to the self-associated and free silanol stretching vibration bands, respectively.

Scheme 1

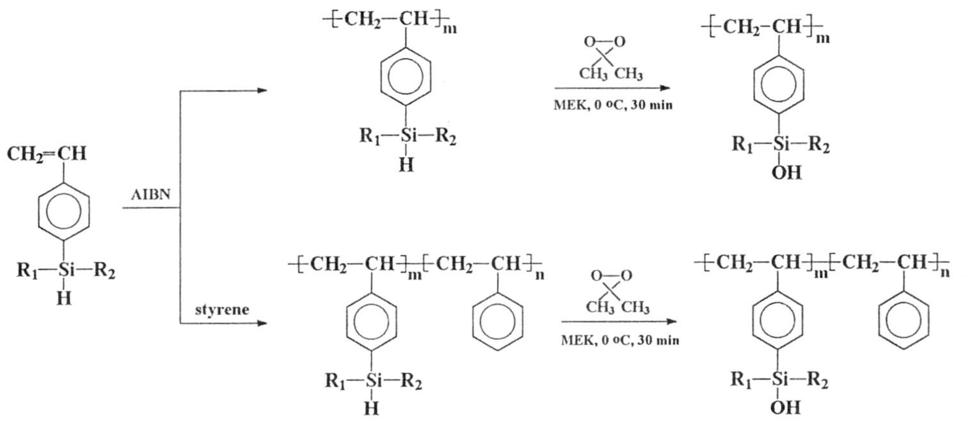

1. $R_1 = CH_3$, $R_2 = CH_3$
2. $R_1 = CH_3$, R_2 = phenyl
3. R_1 = phenyl, R_2 = phenyl

It was found that the stability of these 4-vinylphenyldialkyl/arylsilanol polymer and their styrene copolymers against self-condensation in the solid state depended largely on the substituents bonded directly to the silicon atom. Although 4-vinylphenyldimethylsilanol polymer and styrene copolymers obtained in *situ* showed no tendency for self-condensation in solution at room temperature for several months which is attributed to the formation of the hydrogen bonded complexes between the silanols and the solvent, the copolymers underwent self-condensation by forming insoluble siloxane networks after removal of the solvent when the copolymer contained 18 mole % silanol groups or more. When one of the methyl groups was replaced by the phenyl group, stable 4-vinylphenylmethylphenylsilanol polymer (PVPMPS) and styrene copolymers (ST-VPMPS) were obtained. The increased stability of the polymers against self-condensation in the solid state was attributed to the steric shielding and electron-withdrawing effects of the phenyl substituent bonded directly to the silicon atom. The formation of siloxane crosslinks of PVPMPS at high temperature by the condensation of silanols was characterized by the appearance of a strong Si—O—Si stretching vibration band at 1044 cm^{-1} and a notable reduction in the absorbance of silanol bands in the hydroxyl stretching vibration region (Figure 1C).

The strength of the self-associated silanol hydrogen bonds were reduced considerably with increased steric hindrance of the substituents. The infrared frequency shifts, as a measure of the strength of the silanol self-association between the free and self-associated silanol hydrogen bonds in the silanol stretching vibration region, were

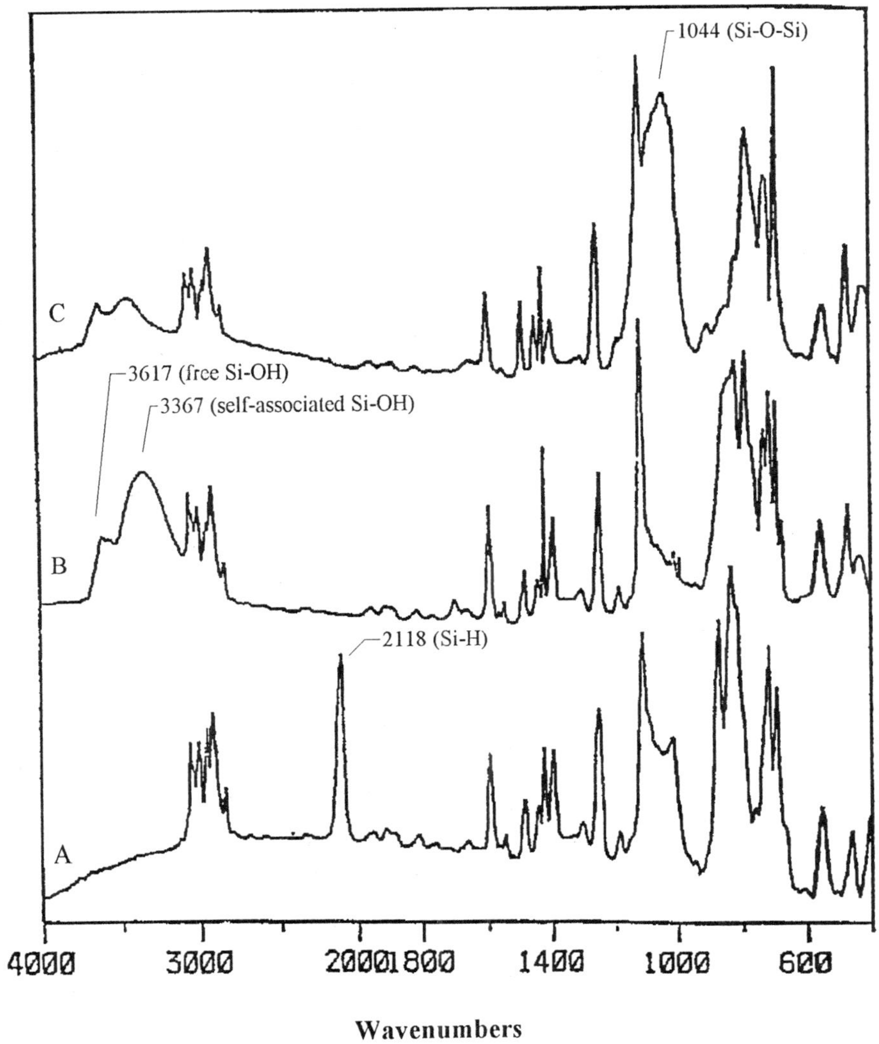

Figure 1. Infrared spectra of (A) poly(4-vinylphenylmethylphenylsilane), (B) poly(4-vinylphenylmethylphenylsilanol) and (C) crosslinked poly(4-vinylphenylmethylphenylsilanol).

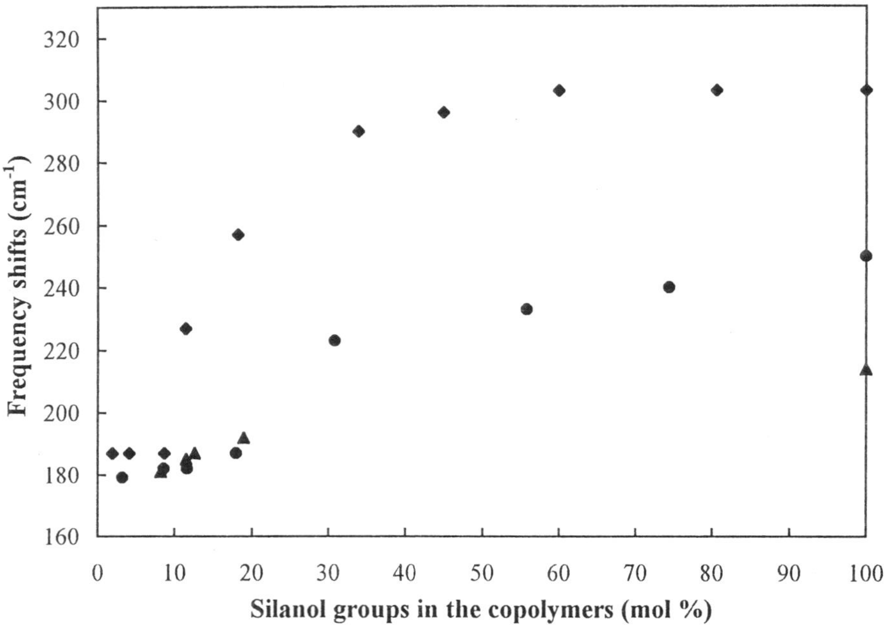

Figure 2. IR frequency shifts in the silanol stretching vibration region. (◆) poly(styrene-co-4-vinylphenyldimethylsilanol) (ST-VPDMS), (●) poly(styrene-co-4-vinylphenylmethylphenylsilanol) (ST-VPMPS) and (▲) poly(styrene-co-4-vinylphenyldisilanol) (ST-VPDPS).

reduced approximately 50 cm^{-1} and 36 cm^{-1} when the methyl groups were replaced by the bulky phenyl groups respectively (Figure 2).

This dependence of the strength of the silanol self-association on the substituents provided an approach for tailor-made miscibility in their binary hydrogen bonded polymer blends.[10] A miscibility window was formed for ST-VPDMS/PBMA blends when the copolymers contained 4 to 34 mol % silanol groups.[11] As a direct result of reduced silanol self-association in ST-VPMPS copolymers, a extended miscibility window was formed when ST-VPMPS copolymers contained 9 to 56 mol % silanol groups.[12]

With controllable strength of the self- and hetero-associated hydrogen, a molecular level penetrated semi-interpenetrating polymer networks (semi-IPNs) of ST-VPDMS with poly(N-vinylpyrrolidone) (PVPr) were prepared by condensation of the self-associated silanols.[13]

Another novel series of inorganic siloxane-based silanol polymers were also synthesized by reacting poly(methylhydrosiloxane) (PMHS) or methylhydrosiloxane-dimethylsiloxane copolymers with a dimethyldioxirane solution in acetone (Scheme 2).[14,15] The infrared spectra showed that the strong stretching vibration band of the Si—H in PMHS at 2166 cm^{-1} totally disappeared after reaction, while a new broad band centered at 3295 cm^{-1}, attributed to the stretching vibration bands of the silanol, appeared in poly(hydroxymethylsiloxane) (PHMS). The formation of the siloxane networks by self-condensation of the silanols in those polymers was characterized by the dramatic decrease of the silanol stretching vibration band at 3295 cm^{-1} and a sharp increase in the absorbance of the siloxane (Si—O—Si) stretching vibration band at 1027 cm^{-1} (Figure 3).

The properties of these siloxane-based silanol polymers depended largely on the silanol composition. Polymer materials obtained by the condensation of the silanols led

Scheme 2

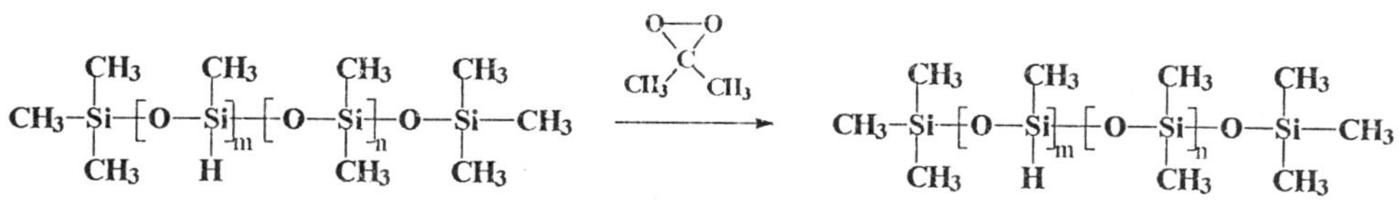

to a wide variety of physical properties ranging from elastomer to thermally stable high T_g crosslinked siloxane polymers. Ceramic-like materials were also obtained in high yield at high temperature under a nitrogen atmosphere.

Hybrid organic-inorganic polymer-polymer composites were prepared by blending with poly(*N*-vinylpyrrolidone) (PVPr), poly(4-vinylpyridine) (PVPy) and poly(ethyl oxazoline) (PEOx). It was found that the properties of these hybrids depended strongly on the preparation conditions.[14]

CONCLUSIONS

A new convenient polymer modification for the conversion of the Si—H to Si—OH by the selective oxidation of the Si—H bond by dimethyldioxirane has been described. The oxyfunctionalization of the silane precursor polymers proceeded rapidly and quantitatively and can be applied to the synthesis of a wide variety of novel silanol polymers with specific properties from the corresponding precursor polymers containing Si—H functional groups. Control over the properties of these silanol polymers, such as reactivity and self-association of silanols, was realized through the placement of different substitute groups bonded directly to the silicon atom and by the variation of silanol composition in a copolymer. These novel silanol polymers with a

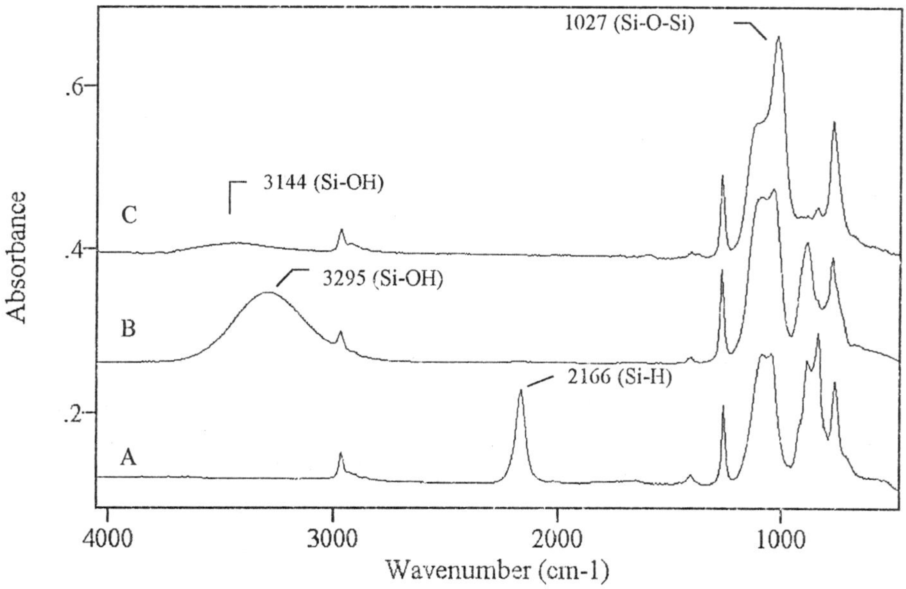

Figure 3. Infrared spectra of (A) poly(methylhydrosiloxane), (B) freshly prepared poly(hydroxymethylsiloxane) and (C) poly(hydroxymethylsiloxane) dried at 80 °C for 24 h.

variety of specific properties can be used as promising materials to prepare miscible polymer blends, semi-IPNs, inorganic-organic polymer-polymer hybrids, preceramic materials, low surface energy coating materials, and contact lenses.

ACKNOWLEDGMENT: The authors are grateful for the financial support provided in part by the National Science Foundation, Division of Materials Research, Polymer Group (Grant No. DMR-9302375).

REFERENCES

1 Noll, W. *Chemistry and Technology of Silicones, Chapter 3*, Academic Press, **1968**.
2 Harris, G. I. *J. Chem. Soc.*, **1963**, 5976.
3 West, R. and Baney, R. H. *J. Inorg. and Nuclear Chem.*, 1958, 7, 297.
4 West, R. and Baney, R. H. *J. Amer. Chem. Soc.*, **1959**, *81*, 6145.
5 Murry, R. W. *Chem. Rev.*, **1989**, *89*, 1187.
6 Adam, W., Curci, R. and Edwards, J. O. *Acc. Chem. Res.*, **1989**, *22*, 205.
7 Adam, W., Mello, R. and Curci, R. *Angew. Chem. Int. Ed. Engl.*, **1990**, *29*, 890.
8 Lu, S. Pearce, E. M. and Kwei, T. K. *Macromolecules*, **1993**, *26*, 3514.
9 Lu, S. Pearce, E. M. and Kwei, T. K. *J. Polym. Sci., Polym. Chem. Ed.,* **1994**, *32*, 2597.
10 Lu, S. Ph.D. Dissertation, Polytechnic University, Brooklyn, New York, **1994**.
11 Lu, S., Pearce, E. M. and Kwei, T. K. *J. Polym. Sci., Polym. Chem.*, **1994,** *32*, 2607.
12 Lu, S., Pearce, E. M. and Kwei, T. K., *J. Polym. Sci., Polym. Chem.*, **1996,** *34*, 3163.
13 Lu, S., Pearce, E. M. and Kwei, T. K., *Polymer*, **1995,** *36*, 2435.
14 Lu, S., Melo, M. M., Zhao, J., Pearce, E. M. and Kwei, T. K. *Macromolecules*, **1995**, *28*, 4908.
15 Pearce, E. M., Kwei, T. K., Gao, Y. M. and S. Lu, unpublished results.

SYNTHESIS AND ION-COUPLING REACTIONS OF TELECHELIC POLY(DIMETHYLSILOXANE) HAVING CYCLIC ONIUM SALT GROUPS

Yasuyuki Tezuka[1], Tomio Iwase[2] and Tomoo Shiomi[2]

[1]Department of Organic and Polymeric Materials,
Tokyo Institute of Technology, O-okayama, Meguro-ku, Tokyo 152
[2]Department of Material Science and Technology,
Nagaoka University of Technology, Kamitomioka, Nagaoka, Niigata 940-21

INTRODUCTION

Uniform size polymers possessing deliberately designed functional end groups are drawing increasing attention as important macromolecular building blocks for the efficient construction of highly controlled macromolecular, and supramolecular structures.[1]

We have recently described on the synthesis and ion-coupling reactions of uniform-size poly(THF) having a series of cyclic onium salt end groups. Among a variety of cyclic onium salt groups, the 5-membered cyclic ammonium (pyrrolidinium) salt is of a particular synthetic interest, since it is stable at room temperature with carboxylate as counter anion, while it undergoes selective and quantitative ring-opening reaction at an elevated temperature to convert the ionic end group to the covalent one. And this unique feature of the pyrrolidinium salt end groups was indeed the key to achieve the efficient macromolecular ion-coupling reactions to produce a variety of branch (star polymers and polymacromonomers) and network (model network) homopolymers as well as graft and network copolymers of precisely controlled architectures.[2-5] We have also reported recently on the synthesis of telechelic polystyrenes having pyrrolidinium salt groups, and subsequent ion-coupling reactions with poly(styrene-co-acrylate salt) to produce well-defined branch and network polystyrenes.[6]

As the extension of the preceding studies, this report describes on the synthesis of poly(dimethylsiloxane), poly(DMS), having pyrrolidinium salt as a single and both end groups (**1c** and **1'c**). The ring-opening reactivity of the end-standing pyrrolidinium groups was then examined with carboxylate as a counter anion, and subsequently the ion-coupling reaction of these telechelic poly(DMS)s was performed with plurifunctional carboxylates, with relevant to those with the poly(THF) and polystyrene telechelics, to produce precisely controlled branch and network poly(DMS) structures.

EXPERIMENTAL

Reagents

Hexamethylcyclotrisiloxane (D_3) (Toshiba Silicone Co.) was distilled over CaH_2. Butyllithium (Nacalai Tesque, 1.6mol/L in hexane) was filtered with a G4 glass filter. A lithium salt of bis[p-(dimethylhydroxysilyl)phenyl]ether was prepared according to the method detailed elsewhere.[7] 3-Trimethylsiloxypropyldimethylchlorosilane, 2., was prepared by the method detailed elsewhere.[6] Tetrahydrofuran (THF) (Nacalai Tesque) was distilled from the blue sodium/benzophenone solution. 1-Methylpyrrolidine (Nacalai Tesque) was distilled over CaH_2. Tosyl chloride and dimethylaminopyridine (Nacalai Tesque) were recrystallized from ethyl acetate. Mono and plurifunctional carboxylates used in the present study were sodium benzoate, sodium terephthalate, sodium and lithium salts of trimesic acid and pyrromelitic acid. They were either purchased or prepared by the standard neutralization method with sodium or lithium hydroxides. Deionized water (chromatography grade, Nacalai Tesque) was used as received.

Synthesis of telechelic poly(DMS) having pyrrolidinium salt groups

Mono- and bifunctionally living poly(DMS) were produced by the method reported previously.[7] The end-capping reaction of the living poly(DMS) was carried out by adding 5-times (for **1**) or 10-times (for **1'**) excess of 2. The reaction mixture was stirred at r.t. for 1h. The product was isolated by pouring the mixture into methanol containing 0.1vol% hydrochloric acid. The oily crude product, **1a** and **1'a**, was recovered after decantation, and purified by the reprecipitation from THF/methanol system, and finally freeze-dried from benzene solution.

Then a weighed amount (15~20g) of **1a** or **1'a** was placed in a 200mL round-bottom flask, and was dissolved in dry CH_2Cl_2 (70mL) under dry N_2 atmosphere. Thereupon a prescribed amount (5 times excess with respect to hydroxyl groups of **1a** and **1'a**) of CH_2Cl_2 solution of tosyl chloride and of dimethylaminopyridine were added by a syringe. The reaction was carried out under stirring at 15°C for 24h. The tosylated product, **1b** and **1'b**, was isolated after evacuating the reaction mixture, and purified by the reprecipitation from THF/methanol system, followed by freeze-drying from benzene solution.

In the final step, a weighed amount of **1b** (10g) and **1'b** (15g) was dissolved in 60mL of benzene and a large excess (20 times with respect to tosyl groups) of 1-methylpyrrolidine was added under dry N_2. The reaction was performed at 50°C for 48h, and the product, **1c** and **1'c**, was isolated after evacuation, and purified by freeze-drying from benzene solution.

Ion-exchange and ring-opening reactions of telechelic poly(DMS) having pyrrolidinium salt groups

A weighed amount of sodium benzoate (twice excess of the end groups) was dissolved into 100mL of deionized water, and a THF solution (1~2mL) of **1c** or **1'c** (0.5g) was added dropwise under vigorous stirring to form a gummy precipitate adhered to the inside wall of the beaker. The stirring was continued at ice-bath temperature for 2h. The precipitated product was then isolated by decantation, linsed by water, and finally freeze-dried from benzene solution. The isolated product was then subjected to the heating treatment at 100°C for 24h in a vacuum descicator.

Ion-coupling reaction of telechelic poly(DMS) having pyrrolidinium salt groups with plurifunctional carboxylates

Into an ice-cooled 100mL aqueous solution containing a prescribed amount of plurifunctional carboxylate compound, a THF solution of an equivalent amount of **1c** and **1'c** (0.5g in 5.0mL of THF) with respect to the ionic groups, was added dropwise by a syringe under vigorous stirring. The resulting turbid solution turned gradually clear, and the precipitate was adhered to the wall of the beaker. The stirring was continued further for 3h, and the product was recovered by decanting the solvent. The product was isolated by freeze-drying from benzene solution, and was subjected to the heating treatment at 120°C for 24h. The gel product obtained from **1'c** was subjected to a Soxhlet extraction with THF for 24h to estimate a gel content.

Measurements

270MHz ^{1}H-NMR spectra were recorded by a JEOL JNM-GX 270 apparatus at 40°C in $CDCl_3$. Chemical shift values were referenced from the residual protonated solvent in $CDCl_3$ (δ =7.30). IR spectra were taken on a SHIMADZU FT-IR 8100 apparatus (40 or 60 scans) using a cast film sample on a NaCl plate from a chloroform solution. GPC measurements were performed using a Tosoh HLC-8020 (an RI detector is pre-installed) with additional UV detector Model UV 8000 at 254nm, and a conductivity detector Model CM 8010. A column system of TSK G3000HXL and TSK G4000HXL was used by a connected mode with THF as eluent at a flow rate of 1.0mL/min.

RESULTS AND DISCUSSION

Synthesis of telechelic poly(DMS) having pyrrolidinium salt groups

We have previously reported on the synthesis of a series of mono- and bifunctional poly(DMS) having a variety of reactive end groups, such as silan (Si-H), vinyl, hydroxyalkyl, dimethylamino and carboxyllic acid groups.[7-12] We have also described already on telechelic poly(DMS) having tosylate end group, **1b** and **1'b**, where the hydrosilation reaction of poly(DMS) having silan end group was performed with allyl alcohol in the presence of Pt/C catalyst, followed by the tosylation reaction with tosyl chloride in the presence of dimethylaminopyridine.[9]

It was found, however, in the course of the present study that the catalytic activity of the commercial Pt/C catalyst for the present hydrosilation reaction was not reproducible from one lot to another. Hence, we have developed an alternative process to produce **1a** and **1'a**, i.e. the direct end-capping reaction of mono- and bifunctionally living poly(DMS) with **2**, which was also applied for the termination reaction of a living polystyrene.[7] The subsequent deprotection of trimethylsilyl-protected hydroxyl groups was achieved under a diluted acidic condition. An elimination reaction of hydroxypropyl end groups was observed under a concentrated acidic medium. The tosylation reaction of hydroxypropyl end groups was performed with tosyl chloride in the presence of dimethylaminopyridine to produce **1b** and **1'b**, and finally the quaternization with 1-methylpyrrolidine was performed in benzene solution to produce **1c** and **1'c** in almost quantitative yields.

^{1}H-NMR spectroscopic analysis shows the signals due to the methylene protons of the α, β and γ-positions in the 3-hydroxypropyldimethylsilyl group for **1a** and **1'a** at 3.66, 1.64 and 0.66ppm, respectively, together with those in the initiator fragments (those of methyl protons at 0.95ppm and of methylene protons at 1.35ppm for **1a**, and of diphenyl protons at 7.00~7.80ppm for **1'a**, respectively) in addition to the overflowed signal at around 0ppm for poly(DMS) main chain.

The subsequent end-group modification can also be monitored by the relevant change in the spectra for **1b**, **1'b** and for **1c**, **1'c**, respectively. Thus after the tosylation treatment, the methylene signal in the tosyl ester group appears at 4.06ppm as a triplet, along with the methyl and phenyl signals due to the tosyl group at 2.49 and at 7.38 and 7.84ppm, respectively. Finally by the quaternization, the signals due to the methyl and methylene groups adjacent to the quaternized nitrogen appear at 3.28 and at 3.64 (exo position) and 3.80ppm (endo position), respectively.(Figure 1, bottom)

Table 1. Synthesis of Telechelic Poly(DMS) Having Hydroxyl (**1a**, **1'a**), Tosyl (**1b**, **1'b**) and Pyrrolidinium Salt Groups (**1c**, **1'c**)

run[a)]	1a, 1'a[b)]					1b, 1'b[c)]		1c, 1'c[d)]	
	initiator[e)] (mmol)	2 (mmol)	yield (%)	$M_n \times 10^{-3}$[f)] calcd	$M_n \times 10^{-3}$[f)] GPC	recovd. yield (%)	$M_n \times 10^{-3}$ (GPC)	$M_n \times 10^{-3}$ (GPC)	end-group[g)] conv. (%)
1	6.11	30.6	93	3.8	3.9	94	4.0	4.1	98
2	4.76	47.2	97	5.4	5.6	97	5.7	5.8	99
3	2.91	28.9	99	8.7	9.0	99	9.1	9.1	97

a) Run 1: **1** (monofunctional poly(DMS)), runs 2,3: **1'** (bifunctional poly(DMS)).
b) D_3; 110mmol, THF; 60mL (run 1) or 100mL (runs 2,3), 1-Methylpyrrolidine; 6.0mL, 25°C, 3h.
c) **1a**: 15-20g, **1'a**: 20g. CH_2Cl_2: 70mL. The reaction with tosyl chloride in the presence of DMAP (OH/tosyl chloride/DMAP = 1/5/5), at 15°C for 24h.
d) **1b**: 10g, **1'b**: 15g. Benzene: 60mL. 1-Methylpyrrolidine (20-times of tosyl groups), at 50°C for 48h. The recovered yield was almost quantitative in each run.
e) Run 1: butyllithium, runs 2,3: a dilithium silanolate having diphenylether group
f) Mn (calcd):
Run 1; D_3/initiator x yield/100 x 222 + 116 (molecular weights of C_4H_9 and $Si(CH_3)_2(CH_2)_3OH$).
Runs 2,3; D_3/initiator x yield/100 x 222 + 402 (molecular weights of $O(C_6H_4)_2$ and 2 x $Si(CH_3)_2(CH_2)_3OH$).
g) By ^{1}H-NMR, based on hydroxyl groups.

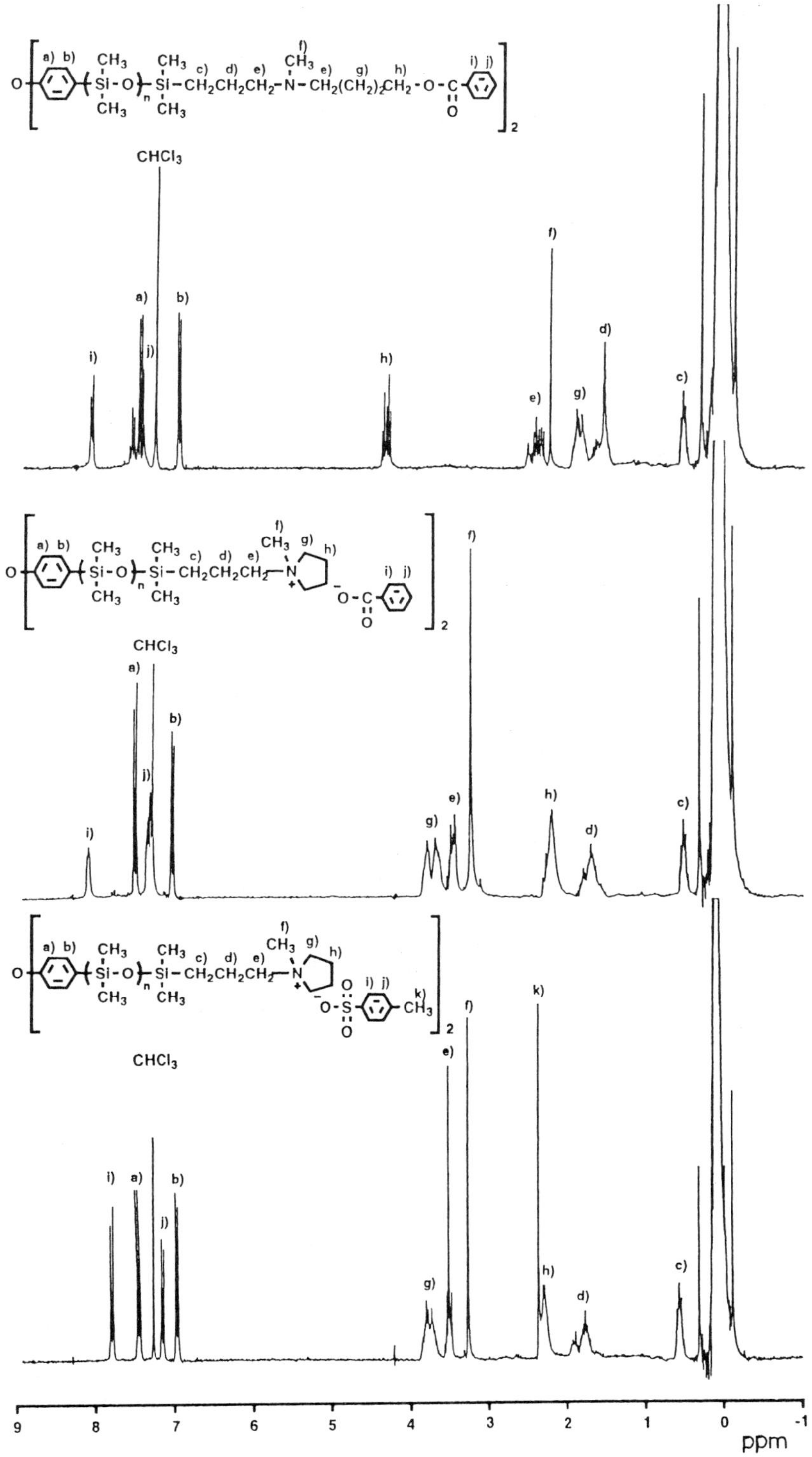

Figure 1. 270MHz ^{1}H-NMR spectra of **1'c** (bottom) and the ion-exchange product with benzoate before (middle) and after (top) the heating treatment at 100°C (Sample: 2 in Table 1, $CDCl_3$, 40°C)

The results of the synthesis of a series of telechelic poly(DMS), **1** and **1'**, are summarized in Table 1. A series of telechelic poly(DMS) of molecular weight up to $9x10^3$ with the narrow molecular weight distribution were produced, in which the observed molecular weights are nearly identical to those estimated from the yield/initiator ratios. The subsequent modification reactions, i.e. the tosylation and quaternization, were found to proceed almost quantitatively to produce **1c** and **1'c** under appropriate reaction conditions.

Ion-exchange and ring-opening reactions of telechelic poly(DMS) having pyrrolidinium salt groups

The ion-exchange reaction was performed by precipitation of **1c** or **1'c** into an aqueous solution containing a slightly excess amount of sodium benzoate. The recovered product was examined by means of ^{1}H-NMR and IR spectroscopic analysis. Typical results are shown in Figures 1 and 2, respectively. The counter anion exchange reaction from a tosylate to a benzoate was confirmed by the appearance of the signals due to benzoate protons at 8.09 and 7.32ppm as seen in Figure 1. IR spectra for the ion-exchange products (Figure 2) showed a carboxylate salt absorption at around 1580cm^{-1}.

The subsequent heating treatment at 100°C caused a selective ring-opening reaction of pyrrolidinium salt by the nucleophilic attack of benzoate counter anion at the endo-position of 5-membered pyrrolidinium ring. Thus the signals due to methyl and methylene protons adjacent to quaternized nitrogen at around 3.0-4.0ppm are removed, and a triplet signal due to ester methylene protons appears at 4.37ppm (Figure 1, top) together with that due to N-methyl protons at 2.26 and N-methylene protons at 2.42ppm, respectively.

IR spectroscopic analysis on the product before and after the heating treatment (Figures 2, B and C, respectively) indicates also that the ring-opening reaction took place in a selective manner. Thus the absorption at around 1580cm^{-1} for carboxylate salt groups is totally removed after the heating treatment, and that at 1730cm^{-1} for covalent ester groups insteadly appears without any noticeable change in other region of the spectrum.

GPC analysis for a series of **1** and **1'** having different end groups was performed with R.I., U.V. and conductivity detectors. The U.V. and conductivity responses show the presence of phenyl groups (by tosylate or benzoate), and ionic groups (by pyrrolidinium) in the product, respectively. The results are collected in Figure 3. It is confirmed therefrom that the molecular weight distribution of the product remained unchanged during the sequential derivatization reactions on the end groups. The U.V. and conductivity responses appear after the tosylation, and after the quaternization reaction, respectively. The conductivity response is totally eliminated after the heating treatment, to indicate the quantitative conversion of ionic groups to covalent ones.

Thus a quantitative ion-exchange reaction was confirmed to take place by a simple precipitation of a THF solution of **1c** and **1'c** into aqueous solution containing a benzoate anion, and the subsequent ring-opening reaction of the pyrrolidinium salt groups was found to proceed by the heating treatment at 100°C.

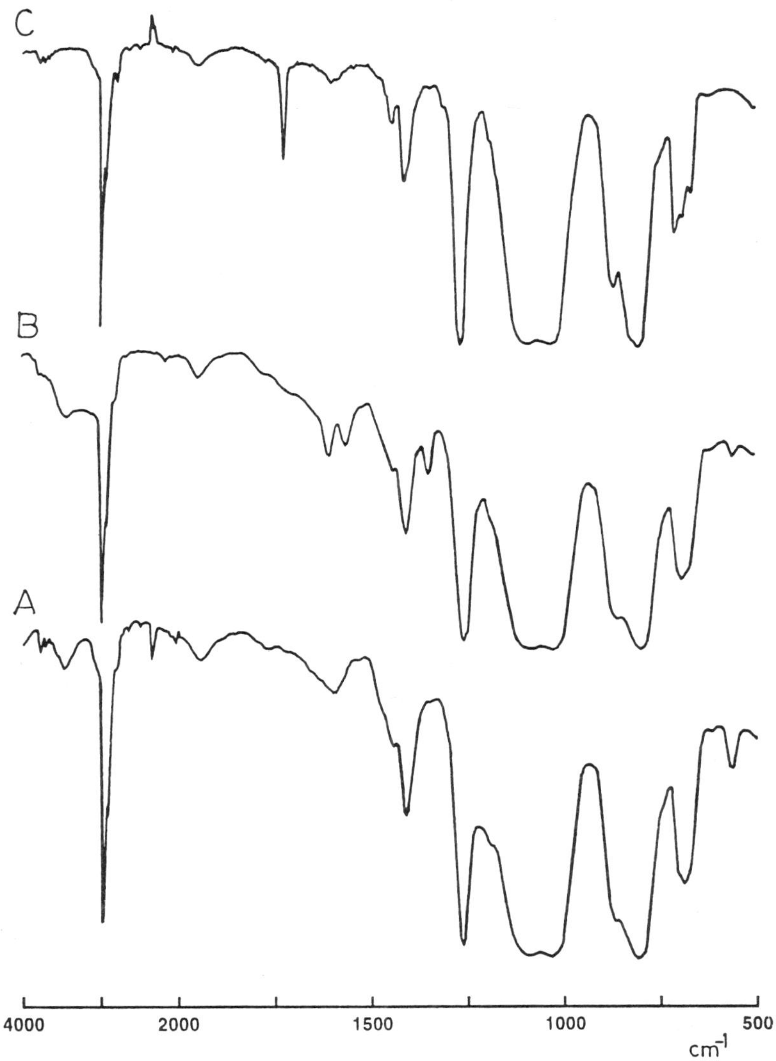

Figure 2. IR spectra of **1c** (A) and the ion-exchange product with benzoate before (B) and after (C) the heating treatment at 100°C (Sample: 1 in Table 1)

Ion-coupling reactions of telechelic poly(DMS) having pyrrolidinium salt groups with plurifunctional carboxylates[13]

The ion coupling reaction was performed in the present study by **1c** and **1'c** with plurifunctional carboxylates to produce poly(DMS)-based well-defined branch and network polymers (Scheme 1). A simple addition of a THF solution of **1c** or **1'c** into an aqueous solution containing an equivalent amount, with respect to the ionic quantity, of plurifunctional carboxylate resulted in the precipitated product adhered to the wall of the beaker. The isolated product was subsequently subjected to the heating treatment at 120°C for 24h.

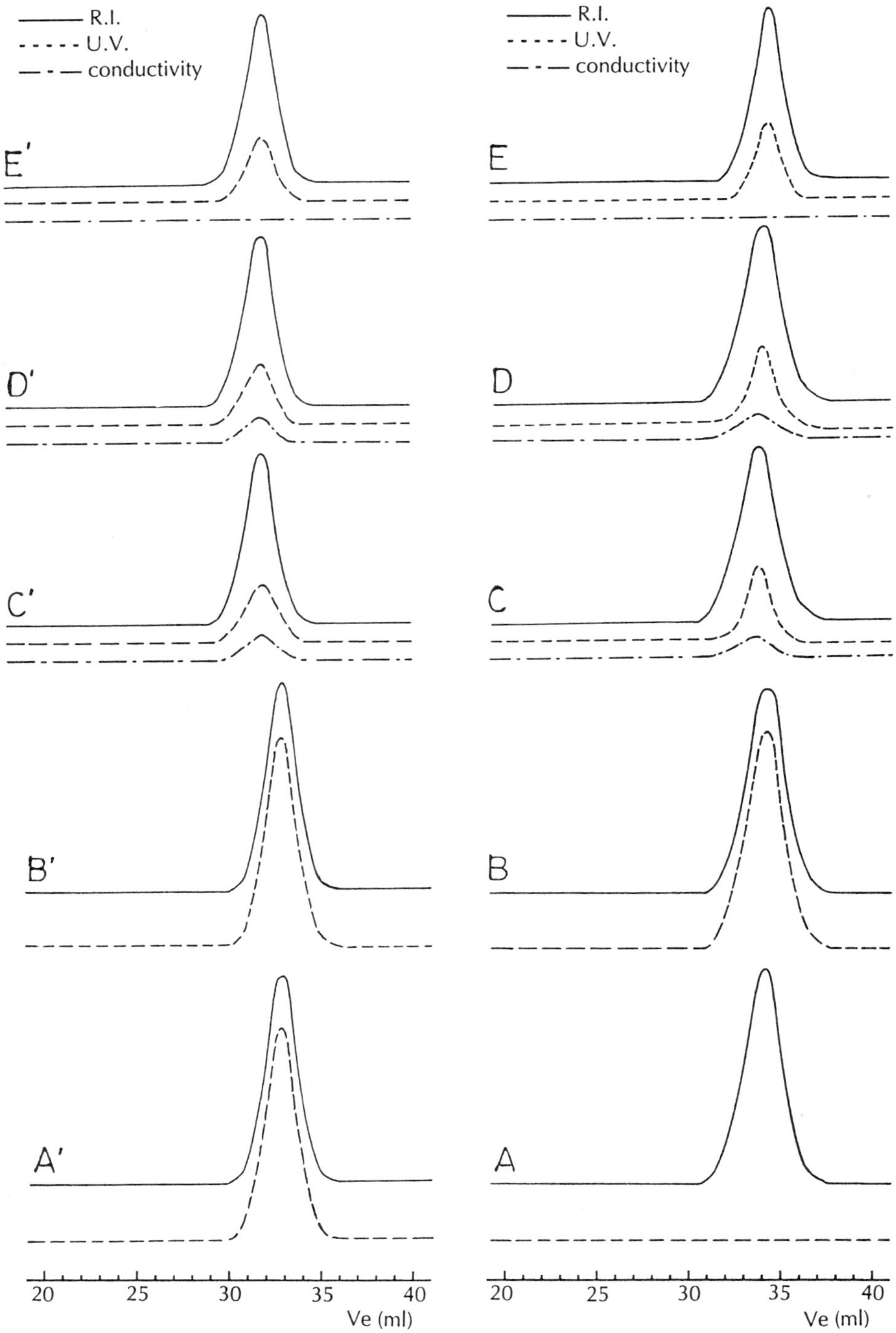

Figure 3. GPC traces of **1a** and **1'a** (A and A'), **1b** and **1'b** (B and B'), **1c** and **1'c** (C and C'), and the ion-exchange product with benzoate before (D and D') and after (E and E') the heating treatment at 100°C (Samples: **1**; 1 1';2 in Table 1)

GPC study on the ion coupling products between monofunctional **1c** and a series of plurifunctional carboxylates showed that the coupling product of **1c** with either bi- or trifunctional carboxylates is unimodal, and the molecular weight of the product is estimated to be twice and three times of the starting **1c**, respectively. The conductivity response is completely eliminated after the heating treatment, indicating the ionic groups were converted to the covalent ones through the nucleophilic attack of carboxylate anion toward the pyrrolidinium ring. On the contrary, the coupling protucts of **1c** with either tetracarboxylate showed no more unimodal distribution. This is consistent with the incomplete ion-exchange reaction with these carboxylates.

The ion couling reaction of bifunctional **1'c** with tri- and tetrafunctional carboxylates was carried out to produce first the ion-exchange, pseudo-network, products crosslinked through the Coulombic interaction, which was found to remain soluble in a good solvent like THF. This unique feature of pseudo-network products allows one to observe the products by solution ^{1}H NMR spectroscopy. The spectra of the ion-exchange products of **1'c** with tri- and tetrafunctional carboxylates demonstrated the nearly quantitative ion-exchange reaction to occur for both systems.

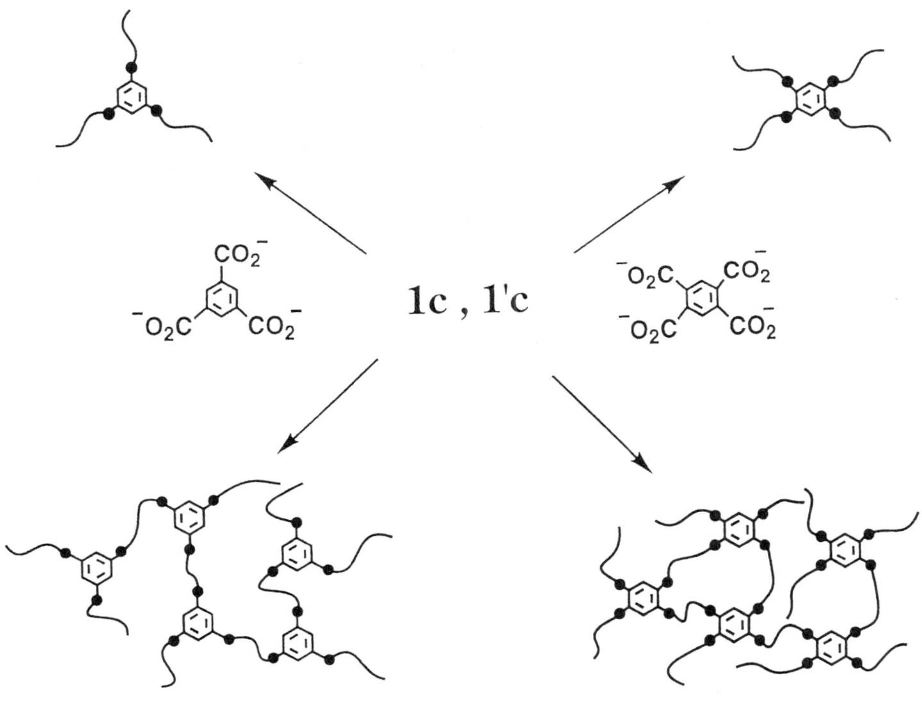

Scheme 1

The subsequent heating treatment of thus obtained ionically-linked, pseudo-network products at 120°C could convert the ionic bond to the covalent one to produce permanent gel products, which were no more soluble but swallen in THF. The ring-opening reaction occurring during the heating was evidenced by IR spectroscopic analysis, where the absorption at 1730cm^{-1} due to the ester groups appears after the heating by eliminating that at 1580cm^{-1} due to carboxylate salt groups without any noticeable change in other spectral area. The network product obtained in the present sysyem are considered as a "model network", in which both the number of the segment at the junction points and the length of the segments between the junction points are precisely defined.

Acknowledgments

This work was supported partly by a grant from the Ministry of Education, Science and Culture, Japan (08455436).

REFERENCES

(1) Tezuka, Y., *Prog. Polym. Sci.*, 1992, *17*, 471.
(2) Tezuka, Y.; Goethals, E.J., *Makromol Chem.*, 1987, *188*, 783.
(3) Tezuka, Y.; Goethals, E.J., *Makromol Chem.*, 1987, *188*, 791.
(4) Tezuka, Y.; Shida, T.; Shiomi, T.; Imai, K.; Goethals, E.J., *Macromolecules*, 1993, *26*, 575.
(5) Tezuka, Y.; Hayashi, S., *Macromolecules*, 1995, *28*, 3038.
(6) Tezuka, Y.; Imai, H.; Shiomi, T., *Macromol. Chem. Phys.*, in press.
(7) Kazama, H.; Tezuka, Y., Imai, K., *Macromolecules*, 1991, *24*, 122.
(8) Tezuka, Y.; Fukushima, A.; Imai, K., *Makromol Chem.*, 1985, *186*, 685.
(9) Kazama, H.; Tezuka, Y., Imai, K., *Polym. J.*, 1987, *19*, 1091.
(10)Kazama, H.; Tezuka, Y., Imai, K., *Makromol Chem.*, 1988, *189*, 985.
(11)Kazama, H.; Tezuka, Y., Imai, K., *Polym.-Plast. Technol. Eng.*, 1989, *28*, 109.
(12)Kazama, H.; Tezuka, Y., Imai, K., *Polym. Bull.*, 1989, *21*, 31.
(13)Tezuka, Y.; Iwase, T.; Shiomi, T., submitted for publication.

AN EMPIRICAL AND MODELING STUDY OF BORON SPECIATION IN SOLUTION WITH A REACTIVE DENDRIMERIC POLYMER

Bryan M. Smith, Paul Todd, and Christopher N. Bowman*
Campus Box 424, Department of Chemical Engineering,
University of Colorado, Boulder, CO 80309-0424
*Corresponding author

ABSTRACT

Developing polymer-assisted ultrafiltration (PAUF) as a viable technology for the removal and concentration of boric acid requires understanding the influence of polymer morphology on boron speciation and chelating effectiveness. To this end, a fifth generation poly(amido amine) dendrimer was persubstituted with borate-specific glucoheptonamide chelating groups and this polymer was used to test a simple partitioning model[1] of boron speciation. The model revises the classic laws of mass action by considering the polymer units in solution to be regions of high local functional group concentration surrounded by a continuous aqueous phase containing no functional groups. Model assumptions and their consequences are frankly discussed. Based on this model, the radius of gyration of the dendrimer is estimated to be an unphysically large 88.5 Å. Compared both to ^{11}B NMR studies of equilibrium boron speciation and to boron breakthrough in a PAUF application, the model tends to underpredict the association between borate an the polymer. This is possibly due to the coulombic interaction between the cationic polyelectrolyte and anionic borate.

INTRODUCTION

This work is motivated by the desire to develop a viable technology for boron removal and concentration from aqueous systems. Such a technology could be applied to the remediation of boron contamination and also in the nuclear power industry, where boron's high neutron absorption cross section is used to control the rates of nuclear reactions. The process under investigation, polymer-assisted ultrafiltration (PAUF), relies on selective and reversible reactions between soluble polymers and the small target molecules to retain the target substance behind an ultrafiltration membrane of molecular weight cut-off suitable to retain the polymer. We have previously demonstrated the feasibility of boron removal and concentration by PAUF[2]. This process has much in common with the use of adsorption resins and often the same adsorption chemistries are employed in both applications. Kinetics of diffusion into and adsorption onto chelating resins are notoriously slow, however, with flow rates typically on the order of a few bed-volumes per hour. In PAUF, diffusional kinetics are not typically limiting, as the polymer

Polymer Modification
Edited by Swift *et al.*, Plenum Press, New York, 1997

regions are now on the order of angstroms, not millimeters; the rate of fluid processing is limited primarily by the surface area of the membrane employed.

PAUF, first suggested by Michaels in 1968[3], has been applied to a wide variety of target compounds. A recent review by Geckeler[4] summarizes much of the work in the field. Many polymeric reagents used in the process have been adapted from other applications (*e.g.* flocculation aids); few have been synthesized for this specific application, though some such chelating polymers have been patented[5]. It is our goal to model the boron PAUF process with special emphasis on the equilibrium speciation of boron in these systems as a means toward rational design of chelating polymers specifically for PAUF.

We have synthesized and tested an example boron-chelating polymer based on a commercially available dendrimeric poly(amido amine) (PAMAM). Dendrimeric chelants offer several advantages over polymers typically used in PAUF. Foremost among these is the reduced viscosity of dendrimer solutions as compared to solutions of linear polymers[6]. This allows the use of higher polymer concentrations than previously feasible (though in the present study we worked at polymer concentrations of less than 5% due to the expense of the starting dendrimer). In this study, the dendrimeric chelant also serves as a convenient, monodisperse polymer with which to test the mathematical model for boron speciation which is derived from the work of Wise and Weber[1].

We present here the initial modeling efforts and experimental determinations of boron speciation in solution with a reactive polymer. Also, the speciation model is applied to the ultrafiltration process and compared with data obtained from PAUF of a synthetic boron-contaminated feed.

BORON CHEMISTRY

A brief review of the relevant complexation chemistry of boric acid is necessary to appreciate both the modeling and experimental efforts described. The reactions of the borate ion with polyhydroxy compounds are well known[7-13] and are shown in Figure 1. 1,3-diols (n=1 in Figure 1) also react with boric acid[14], but the complexes are much less stable than the borate complexes studied here and have therefore been neglected in our modeling. The association constants for borate esters range from around 1 for simple glycols to several thousand for carbohydrates containing longer chains of adjacent hydroxyl groups[10]. The functionalized dendrimer (GP3), a poly(amido amine) persubstituted with D-*glycero*-D-*gulo*-heptonamide residues in place of the terminal amines, is pictured in Figure 2. The association constants for the borate esters and diesters formed in this work are K_1= 6300 M^{-1} and K_2=105 M^{-1} [13].

^{11}B NMR was used to quantify the concentrations of the various boron species in solution. Borate diesters exhibit signals distinct from those of borate esters, and both are distinct from unbound boron. Therefore, with knowledge of the solution pH -- to partition the unbound boron signal into boric acid and the active borate anion -- as well as total boron concentration, the actual species concentrations can be determined.

MODEL DESCRIPTION

We have applied the simple phase partitioning model of Wise and Weber[1] to the borate-dendrimer system. Only a summary of model features will be presented here; readers are referred to the original publication for full details. The model was developed in light of observations that borate diesters (BL_2) were much more prevalent in solutions containing functional polymers and borate than one would expect based on the concentration of functional groups. Since borate esters (BL) form with functional groups on the polymer, the remainder of the polymer chain occupies the local region of a given

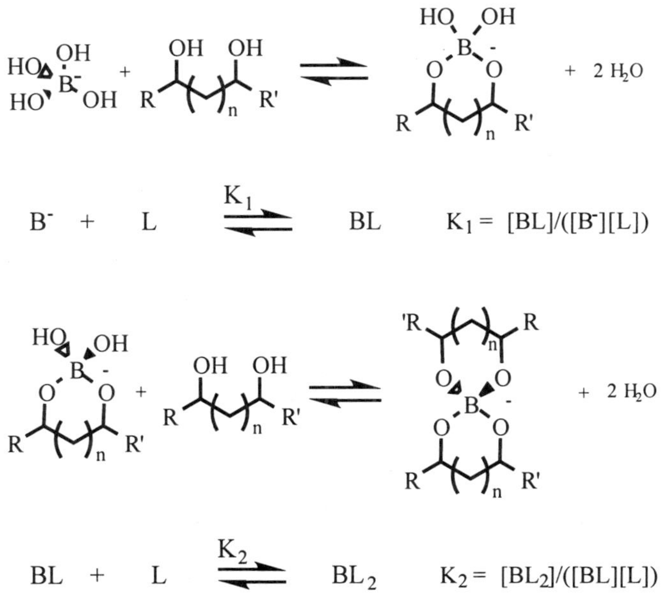

Figure 1. Generalized reactions of borate anion with polyhydroxy compounds (n=0 or 1). K_1 and K_2 are defined in this figure.

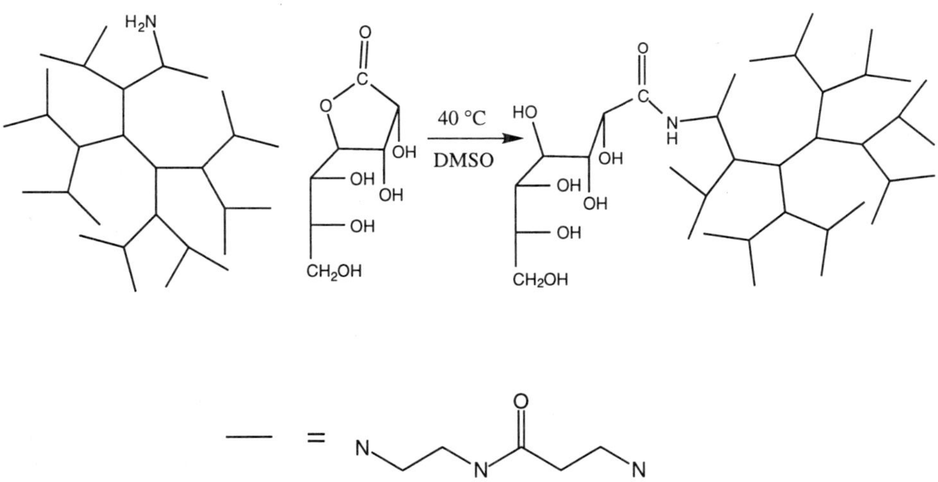

Figure 2. The synthesis of glucoheptonamide-persubstituted dendrimer (GP3) from a diaminoethane core. The dendrimer employed in this report was fifth generation while the one shown here is second generation. Note that for simplicity, only one of the terminal amines is shown explicitly.

borate ester, so the local concentration of functional groups is much higher than that in the bulk solution.

In essence, the model divides a reactive polymer solution into a dispersed polymer-rich phase (phase 1), within which the concentration of functional groups is defined by the polymer morphology and structure, and a solvent-rich phase which contains no functional groups (phase 0). The individual polymer molecules are modeled as spheres of polymer-rich phase stuck at points of an imaginary lattice in solution. If the polymer concentration is sufficiently high, another 'phase' enters the calculations which consists of overlapping polymer-rich spheres (phase 2).

Within either of the polymer-rich phases, borate esters and diesters of the functional groups are assumed to form with the same association constants as observed for the independent functional groups in aqueous solution. The resulting equations simply describe the borate ester association constants as well as mass balances on boron and polymer-bound functional groups. Wise and Weber used the model to estimate association constants for the borate esters formed with the diols in PVA and to predict the gelation of PVA-borate solutions. As we have independently measured the association constants for the borate esters formed in this work, we have used the model to estimate the radius of gyration of the GP3 dendrimer and to predict both the boron speciation in borate/GP3 solutions and the efficacy of PAUF using these functional dendrimers.

The only revision of the model which has been incorporated here is the formal description of the functional group concentration in the polymer rich phases. In our work with the dendrimer, nominally containing 128 terminal functional groups, we calculated the total ligand concentration within the polymer phase 1 to be

$$l_{t,1} = \frac{128}{N_{av} \bullet v_s \bullet (1 - f_p)} \qquad (1)$$

where v_s is the volume of a polymer sphere

$$(v_s = \frac{4\pi R_g^3}{3}) \qquad (2)$$

and R_g is the radius of gyration of the polymer, f_p is the volume fraction of v_s occupied by polymer, and N_{av} is Avagadro's number. Correspondingly, the ligand concentration in phase 2 (where two polymer units overlap) is taken as

$$l_{t,2} = \frac{l_{t,1} \bullet (1 - f_p)}{(1 - 2f_p)} \qquad (3)$$

The model as we have applied it assumes the following:

- The polymer molecules are evenly distributed in solution. In the work presented here, the centers of the macromolecules were assigned to points on a face-centered cubic lattice. Under conditions of high borate and salt concentrations, polymer aggregation and phase segregation of the solutions have been observed, calling into question this assumption of polymer independence under these conditions.
- As noted above, the polymers in solution are assumed to be monodisperse spheres with their attendant functional groups randomly distributed therein. While the fifth generation PAMAM dendrimers used in this study have been shown to be quite spherical, the spatial distribution of (terminal) functional groups is less certain[17].

- The thermodynamics of borate ester formation are assumed to be unchanged for individual functional groups on the polymer relative to the same functional groups unattached, free in aqueous solution. Certainly the enthalpies of formation for the borate esters and diesters should be quite similar; however, the entropic effects of forming internal cycles through borate diesters might be expected to lower the effective equilibrium constant for that reaction (K_2) at high borate concentrations.
- We neglect activity effects and assume that the pH and borate concentrations are the same in all phases. Each GP3 dendrimer contains over one hundred tertiary amines; under the conditions of the ^{11}B NMR experiments reported here (typically pH 4-9), these amine groups would be largely protonated. As discussed below, this coulombic interaction could strongly affect the model predictions.
- In the case of the ultrafiltration studies, the retentate is assumed to be well mixed and the permeate is assumed to be in equilibrium with the retentate (*i.e.* the boron concentration in the permeate is assumed to be the *free* boron concentration in the retentate).

With these assumptions, the distribution of boron species (among boric acid, borate anion, borate esters, and borate diesters) reduces to a function of boron concentration, polymer concentration, radius of gyration and solvated volume fraction of a polymer unit, solution pH, and the association constants for the borate esters of the functional groups employed.

EXPERIMENTAL

GP3 synthesis. Following the method of Aoi *et al.*[18], the ring opening reaction shown in Figure 2 was used to derivatize the dendrimers. In a typical batch, 10.0 g glucoheptonic-γ-lactone (Aldrich) was added to a solution of 1.93 g fifth generation PAMAM dendrimer (MW 28826, Dendritech) in 30 ml dry Me_2SO. The reaction mixture was maintained at 40 ± 1°C for seven hours, then allowed to cool to room temperature. The mixture was diluted in 8 l deionized water and fed, through constant volume diafiltration, into the holding cell of the ultrafiltration device described below. The volume of the polymer solution in the holding cell was reduced to 100 ml, and purification was accomplished by constant volume diafiltration of this solution with more than 6 l deionized water. A sample was dried *in vacuo* for approximately a week prior to ^{13}C NMR and elemental analysis. The product was completely functionalized at the terminal primary amine sites as ^{13}C NMR signals from these groups (at *ca.* 41 ppm in D_2O) were absent in the product. Elemental analysis suggests the presence of approximately 10% residual water in the polymer: calculated for $C_{2158}H_{4064}N_{506}O_{1148}$: C 46.7, H 7.38, N 12.78, O 33.11; found (Huffman Laboratories, Golden, CO): C 42.68, H 7.77, N 11.59, O 39.41.

^{11}B NMR. Samples for ^{11}B NMR analysis were prepared in 1 ml volumetric flasks from stock solutions of boric acid (Fisher, electrophoresis grade) and GP3. GP3 concentration in the stock solution was determined by total organic carbon analysis (Huffman Laboratories, Golden, CO). After pH was adjusted to 6.0-8.5 with HCl or NaOH, samples were diluted to final volume, and final pHs were measured in the flasks with an Orion PerpHect 300 meter and a narrow Cole-Parmer probe (H-05990-30) calibrated at the appropriate temperature. In all cases, the boron concentration was kept below 0.05 M to virtually eliminate polyborate formation. The mole ratio of functional groups to boron ranged from 1.4 to 8.4. All NMR spectra were recorded on a Varian VXR-300S with a boron-free insert. To avoid interference from broad boron background signals, quartz tubes were used for samples with low boron concentrations. Due to the broad and overlapping resonances,

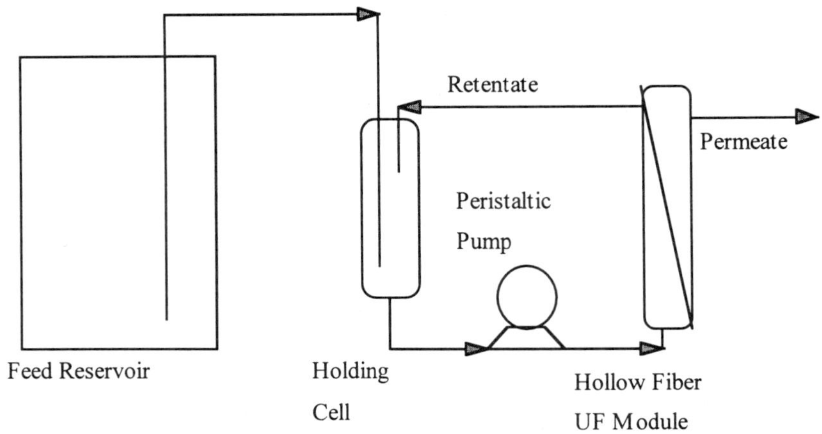

Figure 3. Schematic diagram of the filtration apparatus. The volume in the holding cell was held constant as permeate was replaced by fresh borate-containing feed. Boron concentration and pH were monitored in the permeate.

^{11}B NMR spectra were manually phased and the resonance peaks were fit using the Varian deconvolution routine.

Polymer-Assisted Ultrafiltration of Boric Acid. The Quickstand® (AGT, Needham, MA) filtration apparatus is pictured schematically in Figure 3. The hollow fiber membrane module contained approximately 30 fibers with 0.5 mm internal diameter and had a nominal molecular weight cut-off of 10,000 and a surface area of 0.015 m^2. A pinch clamp in the retentate recycle line was used to supply back pressure to the system. In a typical run, the transmembrane pressure was maintained at 25 psig and the retentate and permeate flow rates were 25 ml/min and 3 ml/min, respectively. Permeate flux remained constant throughout the experiments.

RESULTS AND DISCUSSION

^{11}B NMR. In a dilute polymer solution, the ratio of borate diesters to borate esters is expected to be nearly independent of the polymer concentration and this has been observed by other authors[1, 15, 16]. The local concentration of functional groups is quite high (and constant) in the vicinity of an already formed borate ester because the borate ester is attached to a functional polymer. This local concentration is only dependent on polymer morphology and the fraction of functional groups which are already occupied in borate esters, f. While the model predicts the ratio of diesters to monoesters to be nearly constant as a function of global functional group concentration in dilute solution, accounting for the loss of available functional groups by normalizing that same ratio by $1/f$ should produce a straight line of zero slope (whose value is $K_2 * l_{t,1}$) when $BL_2/(BL \bullet f)$ is plotted versus the global free functional group concentration. This plot is shown in Figure 4, where, though there is considerable scatter in the data, the ratiod concentration of borate diesters to borate monoesters clearly does not tend to zero as one would expect for independent functional groups.

The data in Figure 4 were used to estimate the radius of gyration of the GP3 dendrimer. A least squares fit gives $l_{t,1}$ = 0.075 M and hence (equations 1 and 2) R_g = 88.5 Å. This is much larger than the 27 Å reported for the starting fifth generation

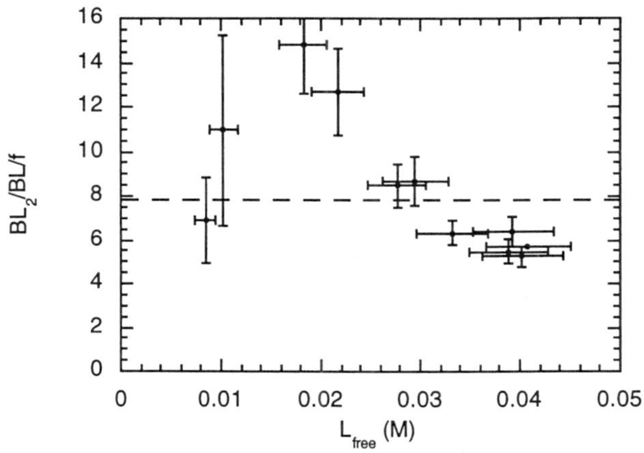

Figure 4. Observed ratio of borate diesters to borate monoesters, scaled by $1/f$, *versus* the global free ligand concentration. Standard errors are shown based on multiple fits to individual ^{11}B NMR spectra. As the free ligand concentration is also calculated (by difference) from the ^{11}B NMR spectra, standard errors are given for this variable as well. The dashed line is the best fit from the model which gives R_g = 88.5 Å. For these samples, f was always between 0.2 and 0.6.

PAMAM dendrimer in methanol[19], and larger also than the 60 Å maximum physical radius of a completely stretched structure, but has been employed consistently. In terms of the model, some of the scatter in Figure 4 can be explained by considering that the radius of gyration of this polyelectrolyte is likely to depend on the ionic strength, pH, and even the concentration of borate diesters (internal cycles); none of these factors were systematically controlled in this study.

In Figure 5 we plot the ratio of borate esters to borate anion as a function of global free ligand concentration. The borate anion concentration was calculated from the free boron concentration (B_f) observed with ^{11}B NMR and the pH:

$$B^- = \frac{B_f}{1+10^{pKa-pH}} \qquad (4)$$

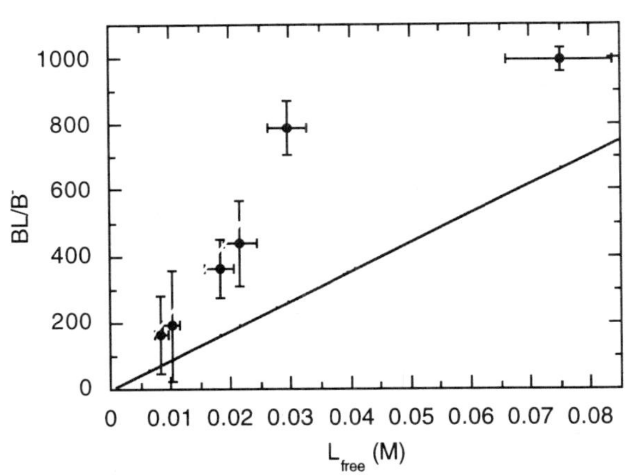

Figure 5. Ratio of borate esters to borate anion as a function of global free ligand concentration. The borate ester concentration was observed directly with ^{11}B NMR, and the borate anion concentration was calculated from the solution pH and the free boron concentration. Standard errors are shown based on multiple fits to individual ^{11}B NMR spectra. As the free ligand concentration is also calculated (by difference) from the ^{11}B NMR spectra, its standard error is shown as well. The model prediction of this ratio, which is identical to the classical law of mass action, is shown as a solid line.

with the pKa of boric acid taken as 9.14. It appears that the model significantly underpredicts the association of borate with the functional groups on the polymer. One possible explanation for this is that coulombic attraction between the cationic polymer and anionic borate may lead to a higher local concentration of borate ions in the region of a polymer unit. It is difficult to test this hypothesis, however, because the GP3 polymer will precipitate in the saline borate solutions required to screen out this coulombic effect. Additionally, if coulombic forces are to be considered, neutralizing the polymer by loading it with borate will surely reduce the apparent association constant K_1. Following the work of Leibler *et al.*[20], accounting for this hypothesized interaction with a Flory-type modification of the apparent association constant may lead to more reasonable predictions of boron speciation.

Polymer-assisted ultrafiltration. The functionalized dendrimer described above was used to remove boron from an aqueous feed stream. With the polymer in the holding cell (see Figure 3), the feed was added at the same rate that the permeate left the system. The differential equation describing the change in total boron concentration in the holding cell as a function of permeate volume is simply

$$V_t \frac{d(B_t)}{d(v)} = B_i - B_p \tag{5}$$

where V_t is the volume of the holding cell, v is the permeate volume, B_t is the total boron concentration, and B_i and B_p are the input and permeate boron concentrations, respectively. The model described above was used to calculate the free boron concentration (B_p) as a function of the total boron concentration at each step as the differential equation was solved numerically.

Figure 6 shows that this polymer is quite effective at removing boron from aqueous streams. A solution containing 1.63 mM B at pH 9.1 was fed into the holding cell which contained 4.755 g GP3 in approximately 100 ml, initially at pH 5.0. Shown in the figure are the model predictions of boron concentration in the permeate, assuming the polymer solution in the holding cell was maintained at pH 7 and, alternatively, at pH 8. Also shown is the permeate concentration predicted in the absence of polymer, an exponentially asymptotic approach to the feed concentration. It is clear that although the model

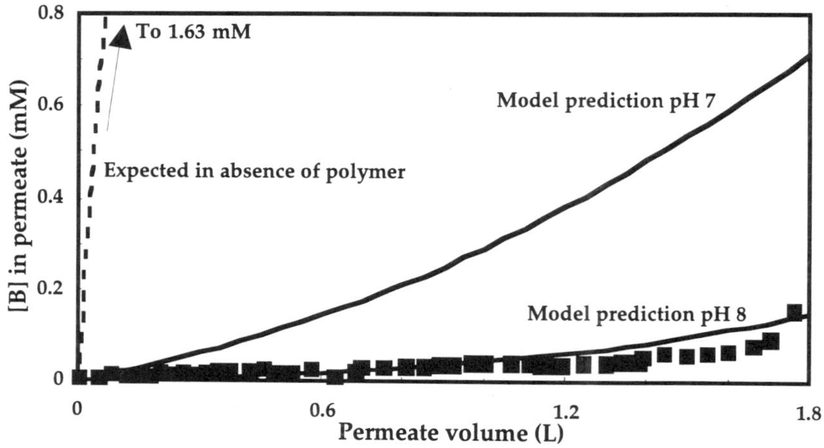

Figure 6. Concentration of boron in the permeate (solid squares) as a function of permeate volume as the 1.63 mM boric acid feed (pH 9) was diafiltered through the holding cell containing 100 ml water and 4.755 g GP3 dendrimer. The permeate pH rose from 5.02 at the beginning of the run to 7.75 at the end. Model parameters: Rg, 88.5 Å; K_1, 6300; K_2,105; Mw, 55471; fp, 0.024.

prediction of boron concentration in the permeate roughly approximates the observations, the model again predicts less boron binding than is actually observed.

CONCLUSIONS

A smiple phase partitioning model for chelating polymers, based on the image of polymer units as local regions of high functional group concentration, was used to estimate the radius of gyration of a borate-specific dendrimeric chelant and to predict the boron speciation in aqueous solutions with this polymer. While the model appears to underestimate the associations between borate and the chelating groups on the polymer, the estimation of radius of gyration is not unreasonable. The boron-speciation model was able to more closely predict the performance of this dendrimer in the polymer-assisted ultrafiltration of boric acid.

ACKNOWLEDGMENTS

Funding for this research was provided by the Center for Separations using Thin Films, the Camille Dreyfus Teacher-Scholar Program and an NSF Presidential Faculty Fellowship for CNB (CTS-9453369).

REFERENCES

1. Wise, E.T. and S.G. Weber, *A simple partitioning model for reversibly cross-linked polymers and application to the poly(vinyl alcohol)/borate system ("slime").* Macromolecules, 1995. **28**(24): p. 8321-8327.
2. Smith, B.M., P. Todd, and C.N. Bowman, *Boron removal by polymer-assisted ultrafiltration.* Separation Science and Technology, 1995. **30**(20): p. 3849-3859.
3. Michaels, A.S., *Ultrafiltration*, in *Advances in separation and purification*, E.S. Perry, Editor. 1968, John Wiley: New York.
4. Geckeler, K.E. and K. Volchek, *Removal of hazardous substances from water using ultrafiltration in conjunction with souble polymers.* Environmental Science and Technology, 1996. **30**(3): p. 725-734.
5. Grinstead, R.R., *Process for removing boron ions from aqueous solutions*, U.S. Patent 4,755,298, 1988.
6. Mourey, T.H., *et al.*, *Unique behavior of dendritic macromolecules: intrinsic viscosity of polyether dendrimers.* Macromolecules, 1992. **25**(9): p. 2401-2406.
7. Kennedy, G.R. and M.J. How, *The interaction of sugars with borate: an N.M.R. spectoscopic study.* Carbohydrate Research, 1973. **28**: p. 13-19.
8. Makkee, M., A.P.G. Kieboom, and H. van Bekkum, *Studies on borate esters III. borate esters of D-mannitol, D-glucitol, D-fructose and D-glucose in water.* Recl. Trav. Chim. Pays-Bas, 1985. **104**(9): p. 230-235.
9. van Duin, M., *et al.*, *Studies on borate esters I. the pH dependence of the stability of esters of boric acid and borate in aqueous medium as studied by ^{11}B NMR.* Tetrahedron, 1984. **40**(15): p. 2901-2911.
10. van Duin, M., *et al.*, *Studies on borate esters II. Structure and stability of borate esters of polyhydroxycarboxylates and related polyols in aqueous media as studied by ^{11}B NMR.* Tetrahedron, 1985. **41**: p. 3411-3421.
11. van Duin, M., *et al.*, *Studies on borate esters IV. Structural analysis of borate esters of polyhydroxycarboxylates in water using ^{13}C and ^{1}H NMR spectroscopy.* Recl. Trav. Chim. Pays-Bas, 1986. **105**(11): p. 488-493.

12. Bell, C.F., R.D. Beauchamp, and E.L. Short, *Borate complexation with pentitols: a study by ^{11}B-N.M.R. spectroscopy and MNDO semi-emperical LCAO-MO calculations.* Carbohydrate Research, 1989. **185**: p. 39-50.
13. Smith, B.M., Doctoral Thesis, 1997, University of Colorado, Boulder.
14. Bachelier, N. and J.-F. Verchere, *Formation of neutral complexes of boric acid with 1,3-diols in organic solvents and in aqueous solution.* Polyhedron, 1995. **14**(13): p. 2009-2017.
15. Pezron, E., *et al.*, *Complex formation in polymer-ion solutions. 1. polymer concentration effects.* Macromolecules, 1988. **22**(3): p. 1169-1174.
16. Sinton, S.W., *Complexation chemistry of sodium borate with poly(vinyl alcohol) and small diols. A ^{11}B NMR study.* Macromolecules, 1987. **20**(10): p. 2430-2441.
17. Murat, M. and G.S. Grest, *Molecular dynamics study of dendrimer molecules in solvents of varying quality.* Macromolecules, 1996. **29**(4): p. 1278-1285.
18. Aoi, K., K. Itoh, and M. Okada, *Globular carbohydrate macromoecular "sugar balls". 1. Synthesis of novel sugar-persubstituted poly(amido amine) dendrimers.* Macromolecules, 1995. **28**(15): p. 5391-5393.
19. Dendritech Inc., *Dendritech, exclusive source of starburst® dendrimers*, 1995: Midland, Michigan.
20. Leibler, L., E. Pezron, and P.A. Pincus, *Viscosity behaviour of polymer solutions in the presence of complexing ions.* Polymer, 1988. **29**: p. 1105-1109.

AUTHOR INDEX

SUBJECT INDEX